P9-AOX-536

The ENERGY

Non Crisis

by

Lindsey Williams
Chaplain to the Trans-Alaska Oil Pipeline

As Told to Dr. Clifford Wilson
Author of *Crash Go The Chariots*
And 25 Other Books

Copyright © 1980

First Edition—July, 1980
Second Edition—1st Printing October, 1980
2nd Printing March, 1981
3rd Printing October, 1981
4th Printing May, 1982
5th Printing June, 1985
6th Printing October, 1985
7th Printing December, 1988

Published by
Worth Publishing Co.

ISBN 0-89051-068-7

ALL RIGHTS RESERVED

No part of this publication may be reproduced, stored in a
retrieval system, or transmitted in any form or by any means—
electronic, mechanical, photocopy, recording, or otherwise—
without the express prior permission of Worth Publishing Company, with the exception of brief excerpts in magazine articles
and/or reviews.

Printed in the United States of America

About the Author

Lindsey Williams, who has been an ordained Baptist minister for 28 years, went to Alaska in 1971 as a missionary. The Transalaska oil pipeline began its construction phase in 1974, and because of Mr. Williams' love for his country and concern for the spiritual welfare of the "pipeliners," he volunteered to serve as Chaplain on the pipeline, with the subsequent full support of the Alyeska Pipeline Company.

Because of the executive status accorded to him as Chaplain, he was given access to the information that is documented in this book.

After numerous public speaking engagements in the western states, certain government officials and concerned individuals urged Mr. Williams to put into print what he saw and heard, stating that they felt this information was vital to national security. Mr. Williams firmly believes that whoever controls energy controls the economy. Thus, *The Energy Non-Crisis*.

Because of the outstanding public response that has been generated by this book, Lindsey Williams is in great demand for speaking engagements, radio, and TV shows. He may be contacted at P.O. Box 7, Kasilof, Alaska 99610, phone (907) 262-4000.

(Addition to the fourth printing of the second edition.)

Order Additional Copies of

THE ENERGY NON-CRISIS

1 Book, $6.00
(Add $1.00 for postage & handling.)

DISCOUNT FOR VOLUME PURCHASES

12 Books, $ 50.00 Postpaid (30% Discount)
100 Books, $360.00 Postpaid (40% Discount)

QUANTITY _____   TOTAL $_____

Name _____

Address _____

City/State/Zip _____

Telephone (_____)_____

MAKE CHECK OR MONEY ORDER PAYABLE TO
WORTH PUBLISHING COMPANY
PAYMENT MUST BE IN AMERICAN FUNDS OR EQUIVALENT

Mail To:
Worth Publishing Co.
P. O. Box 7
Kasilof, AK 99610

Thank you for your order.

Mail To:

Worth Publishing Co.
P. O. Box 7
Kasilof, AK 99610

	PRICE
SYNDROME OF CONTROL	
1 Copy @ $6.00 + $1.00 postage & handling	$_____
12 Books, $50.00 Postpaid	_____
100 Books, $360.00 Postpaid	_____
TO SEDUCE A NATION	
1 Copy @ $6.00 + $1.00 postage & handling	$_____
12 Books, $50.00 Postpaid	_____
100 Books, $360.00 Postpaid	_____
WHERE'S THE FOOD?	
1 Copy @ $6.00 + $1.00 postage & handling	$_____
12 Books, $50.00 Postpaid	_____
100 Books, $360.00 Postpaid	_____
CASSETTE TAPES, $5.50 each Postpaid	
____ Syndrome of Control	$_____
____ America in the Bible	_____
____ The Energy Non-Crisis	_____
____ Political—Monetary 666	_____
____ Where's The Food?	_____
VIDEOCASSETTES, $45.00 each	
____ Torn From the Land	$_____
____ The International Financiers	_____
____ The Energy Non-Crisis (53 min.)	_____
____ To Seduce A Nation	_____
TOTAL	$_____

Name _____

Address _____

City/State/Zip _____

Telephone (_____)_____

MAKE CHECK OR MONEY ORDER PAYABLE TO
WORTH PUBLISHING COMPANY
PAYMENT MUST BE IN AMERICAN FUNDS OR EQUIVALENT

Thank you for your order.

Our President has stated that our energy problem is the equivalent of war. Yet he has embraced policies that have continually discouraged and hampered the development of our oil industry.

Nearly ten years ago President Nixon warned of a pending energy shortage unless our domestic production be drastically increased, but Congress insisted on restrictive price controls.

Congress has been urged—and sometimes threatened—by special interest groups to take a negative stance on energy production, but they have miserably failed to take proper action to increase our domestic production. In fact, as you read this book you must come to the realization that energy production has been fiercely stifled by *"Government Bureaucracy,"* and Congress has sat on its collective hands.

You, the reader, will be left to make your own conclusions as to why this set of facts and circumstances conflict many times with what we have been told by the news media—which is fed *its* information by Government Agencies and Departments.

It is with great pride and pleasure that I endorse this manuscript and compliment the authors for taking time to do the research and make it available to all of us.

Hugh M. Chance
Former Senator of
The State of Colorado

March 19, 1980

Table of Contents

CASSETTE TAPES

VIDEO TAPES

Available By
Lindsey Williams

WRITE FOR LISTING TO

WORTH PUBLISHING CO.
P. O. Box 7
Kasilof, AK 99610

CASSETTE TAPES, $5.50 each Postpaid

"Syndrome of Control"

"America in the Bible."

"The Energy Non-Crisis"

"Political—Monetary 666"

VIDEOCASSETTES, $45.00 each Postpaid

"Torn From the Land"

"The International Financiers"

"The Energy Non-Crisis"

"To Seduce A Nation"

PRICES SUBJECT TO CHANGE WITHOUT NOTICE

A BARREL (42 GALLONS) OF PRUDHOE BAY CRUDE OIL IN $$$

West coast price paid by refinery	$32.50	Approximate averages,
Pipeline tariff from Prudhoe Bay to Valdez	− 2.42	1st quarter, 1981.
Tankering costs from Valdez to West Coast	− .70	(Figures taken from a speech delivered
Wellhead price	$29.38	by Mr. Harold Champagne of ARCO to
		the Dallas chapter of the Society of
Wellhead price (on which taxes are levied)	$29.38	Petroleum Engineers Wives Club)
1/8 royalty (state of Alaska)*	− 3.67	
Severance tax (state of Alaska)*	− 3.50	Absolutely no profit
"Windfall Profits" tax (federal excise tax)*	−11.46	has been earned at
Operating costs (getting oil out of ground)*	− 0.50	this point.
Pretax profit	$10.25	
Gross profit before income tax	$10.25	
Alaska state income tax (9.4%)*	− .96	
Federal income tax (48%)*	− 4.47	
Net profit to producer	$ 4.82 (11¢ per gal.)	

Summary (totals)

		%
Consumer (refinery)	$32.50	100
Costs (pipeline, tankering & operating)	3.62	11
Alaska state tax	8.13	25
Federal tax (including "windfall profits" tax)	15.93	49
Oil company	4.82	15

NOTE
Combined state and
federal tax / 74%

Total value per day	$48,750,000
*Total taxes per day	36,090,000

(To verify tax figures, write Department of Revenue, 201 E. 9th Ave., Anchorage, AK 99501
and the Department of Energy, 1000 Independence Avenue SW, Washington, D.C. 20585)

Chapter 1

The Great Oil Deception

There is no true energy crisis. There never has been an energy crisis ... except as it has been produced by the Federal government for the purpose of controlling the American people. That's a rather dramatic statement to make, isn't it? But you see, at one time I too thought there was an energy crisis. After all, that was what I had been told by the news media and by the Federal government. I thought we were running out of crude oil and natural gas. Then I heard, I saw, and I experienced what I am about to write. I soon came to realize that *there is no energy crisis.* There is no need for America to go cold or for gas to be rationed. We shall verify these statements as we provide the facts for you. You might be surprised to find that we will also show why the price of gas will remain high, and in fact will go higher than it is now.

You've read about the controversy. You've heard the statements, the claims, the counterclaims. You've read about the problems of environmental protection, such as the need to protect birds whose species are becoming extinct. What you haven't heard is that $2 million dollars was spent to go around the nest of one species. On your property, you'd have moved the nest—not so on the Alaska Pipeline. Not true? Questionable? We'll give you the facts.

You've read about the objections of the native Alaskans whose territory is being exploited by those giant corporations that can never be satisfied. You've heard about the excessive profits made by the oil companies. But you haven't heard about the incredible regulations that forced the costs of the Trans-Alaska oil pipeline up from a projected $2 billion dollars to beyond $12 billion dollars. We'll tell you more about that.

I became convinced of the fact that there is no energy crisis when a Senator visited me on the Pipeline. As well as being a former Senator of the State of Colorado, he is also an outstanding Christian gentleman. He came to the Pipeline at my invitation, to speak in the work camps for which I was responsible as Chaplain, on the northern sector of the Trans-Alaska Oil Pipeline.

While I was there I arranged for him to have a tour of the Prudhoe Bay facility. The Senator was shown everything he wanted to see, and he was told everything he wanted to know. The Senator was given information by a number of highly-placed responsible executives with Atlantic Richfield, and these were cooperative with him at all times. He especially gained information from one particular official whom we shall call Mr. X, because of the obvious need to protect his anonymity.

After the Senator had talked at length with Mr. X, we came back to my dormitory room at Pump Station No. 1 and sat down. The Senator said to me, "Lindsey, I can hardly believe what I have seen and heard today."

I waited to see what it was that was so startling. Remember, as yet I had no inkling that there was, in fact, no true energy crisis.

The Senator was very serious. He was obviously disturbed. He looked up at me as he said, "Lindsey, I was in the Senate of the State of Colorado when the Federal briefers came to inform us as to why there is an energy crisis. Lindsey, what I have heard and seen today, compared with what I was told in the Senate of the State of Colorado, makes me realize that almost everything I was told by those Federal briefers was a downright lie!''

At that point the Senator asked if I could arrange for another interview with Mr. X on the following day. I did arrange for that interview, and the Senator and Mr. X sat in Mr. X's office. I was allowed to be present as the Senator asked question after question after question.

The Senator's first question was, ''Mr. X, how much crude oil is there under the North Slope of Alaska, in your estimation?''

Mr. X answered, ''In my estimation, from the seismographic work and the drillings we have already done, I am convinced that there is as much oil under the North Slope of Alaska as there is in all of Saudi Arabia.''

The Senator's next question was perhaps an obvious one. ''Why isn't this oil being produced, if there is an oil crisis?'' He went on to point out that private enterprise has always come to the rescue of the American people when there have been times of need.

Mr. X then made the startling observations that the Federal government and the State government of Alaska had allowed *only one* pool of oil on the North

Slope of Alaska to be developed.

The Senator then asked, "Mr. X, do you think that there are numerous pools of oil under the North Slope of Alaska?"

Mr. X replied, "Senator, the government has allowed us to develop only one 100-square-mile area of this vast North Slope. There are many, many 100-square-mile areas under the North Slope of Alaska which contain oil. There are many pools of oil under the North Slope of Alaska."

The Senator then asked, "Mr. X, what do you think the Federal government is out to do—what do you *really* think the government has as its *ultimate goal* in this business?"

Mr. X's answer was highly controversial in its implications. He stated, "I *personally* believe that the Federal government is out to declare American Telephone and Telegraph a monopoly. In so doing they will be able to divide the company and to break the back of the largest private enterprise on the face of the earth. Secondly, they want to nationalize the oil companies. I believe that these two objectives merge." As Mr. X continued to elaborate his point of view, it became clear that the objectives, as he saw them, were of dramatic import for the economic welfare of this country and indeed for the whole world.

The Senator asked one last question, "Mr. X, if what you say is true, then why don't you as oil companies tell the American people the truth and warn them?"

"Senator," Mr. X replied, "we don't dare tell the American people the truth because there are so many laws already passed and regulations on the books that if the government decided to impose them all on us and enforce them, they could put us into bankruptcy within six months."

In light of what Mr. X stated in that conversation with the Senator, it would seem that the stakes are even bigger than money. They would involve power and domination—initially under the guise of government ownership and control of only the essential commodities and services, but then progressively beyond that. We would call it socialism. Others would give it different names. In the light of Mr. X's statements, that is the direction in which America is being led post-haste today. This book is an attempt to awaken the public to the facts before it is too late.

Mr. X is a man whose observations must be taken seriously. He was one of the numerous executives with Atlantic Richfield who was given the responsibility of developing the entire East side of the oil field at Prudhoe Bay. His credibility cannot be denied. Mr. X has developed numerous oil fields for Atlantic Richfield throughout the world and has built numerous refineries. He is an expert in this field.

So far we have given you just a few side observations. But there is more. Much more. We have a story that must be told. There are tremendously important matters involved—matters of principle and the concepts highly important to our whole way of life. They involve politics, economics, and our American way of life.

Keep reading!

Chapter 2

Establishing Credibility

In this book we will at first give only observations and not opinions. This will set the stage for others to arrive at informed conclusions. At the summation of the book, however, we will allow ourselves the luxury of expressing some opinions—where they are clearly justified by the observations we have made. My primary objective is to report observations, factual material that often could not otherwise be known. Some of it is startling and highly controversial, in that it relates to decisions of policy and high prices, and it is certainly highly relevant to America's national interests—which, of course, makes it of dramatic importance to the rest of the world, as well.

Such statements might seem to be sweeping—some people will even regard them as outrageous. Nevertheless, they're made with the knowledge that they are accurate and vital, and with the conviction that they ought to be told. That being so, why should they not be taken seriously? Plenty of people have said there is no true energy crisis, but almost always they make those statements based on rumors and hearsay; seldom are they able to back up their statements with solid facts.

That is where this book is different. At the risk of being misunderstood, it is necessary to demonstrate that the observations that follow come from a reputable and unprejudiced witness. Credibility must necessarily be established.

Probably it should first be stated that I am an ordained Baptist Pastor and have been a minister of the Gospel of Jesus Christ for over 20 years. In fact, that is an important reason why I received access to the information presented in this book—first, because I was a Chaplain to the Trans-Alaska Oil Pipeline; second, because that position gave me executive status, and with it access to a great deal of information that would not be available to the "man on the street." On the other hand, I have not revealed anything of a confidential nature. At no point have I been asked to withhold any of the information that is presented in this book. Officials have talked to me

freely, have shown me technical data, and
have explained the intricacies of their
highly-complex operations at every point
that I showed interest. They have never
embarrassed me because of my original
lack of knowledge about their field, but
have been courteous and have led me to
an in-depth understanding of the workings of the total oil field. They carefully
went through all sorts of detail when I
was there with the Senator, explain-
ing from their own model of the field
where the wells were, what their depth
was, how much oil was available in
the areas where they had drilled—and
so much more. I saw their seismographic
information, discussed with them their
ideas as to how much oil was at one
point and another, and asked all those
questions which might be asked by any
intelligent observer with an interest in
this, the greatest project ever undertaken
by private enterprise in the whole of the
history of the world.

I learned that there were two ways to
know how much oil was in a particular
area—by seismographics and by actually
drilling right into the oil field itself. I
had free access to the jobs where the
men were working, even on the rigs
themselves, and I was able to watch them
drilling. Later we shall see that this is
highly relevant to some of the important

conclusions that many will draw after reading this book.

I always had access to the technical data in the offices; it was made readily available to me. It was open and above-board; there was no question of confidentiality being breached, and indeed after my eyes had been opened to the fact of a *non-energy crisis,* the cooperation was even greater than it had been before. Many officials are likewise concerned at what the government was and is doing to oil companies, and to the supply of oil to the people of America.

We headed our chapter with a reference to credibility. Another aspect that must be stated is that I did not have the proverbial ax to grind, either with the oil companies or with the government. The oil companies never asked me to be a Chaplain on the Trans-Alaska Oil Pipeline—indeed, the opposite is true. It took six months of pleading my case, of being shuttled from official to official, of being given a regular runaround, before I managed to obtain status as a Chaplain. Eventually, the personnel relations official with Alyeska Pipeline Service Company, Mr. R. H. King, gave me authorization to work directly under the auspices of Alyeska Pipeline Service Company as a Chaplain. The company that was formed by a consortium of nine major oil com-

panies of America was called Alyeska
Pipeline Service Company. The Pipeline
officials allowed me on the Pipeline as
Chaplain with considerable reluctance. I
was the first Chaplain appointed, and I
was the only Chaplain who stayed right
through the entire project. The original
thinking of the officials was that a Chaplain would be out of place with the type
of personnel associated with the rough
and tough oil industry. After being on
the Pipeline for a period of time, they
realized the value of having a Chaplain.
Mr. R. H. King, himself, the Personnel
Relations man from Alyeska who appointed me, acknowledged that I was sav-
ing the company thousands of dollars
every week through my counseling and
the general atmosphere I was creating in
the camps.

At that point, because the company
could not pay me, due to the original
agreement at the time of my appoint-
ment, they decided to give me executive
status. This meant that I had highly val-
ued privileges, as well as access to data
which was not classified confidential, but
nevertheless was highly important in the
national interest. In lieu of monetary
payment, they decided to compensate me
by giving me executive privileges.

In going to the Pipeline, I had no in-
tentions of being (or becoming) involved

in political issues. Indeed, my whole motivation was to help the men spiritually. I totally believe in my work as a Baptist Minister, and here was a tremendous challenge. I have always been ready to see a challenge and to fight for what I believe. When I found that the idea of a Chaplain to the Pipeline was almost anathema to the Pipeline officials, it made me realize even more than ever before that this was a real mission field. I regarded those men on the Pipeline as sheep without a shepherd, and simply stated, my heart went out to them.

It was only after my eyes were opened at the time of the discussions with the Senator and Mr. X that I was led into a totally different understanding of a troublesome situation—which I realized must be faced and presented to the American people. Hence this book.

I submit that my credibility is established. I worked on the Pipeline for two and one-half years. I was not paid by either the oil company or any government agency for all of that time, and I believe that I am entitled to claim in sincerity that I had no bias and no particular pleading. I was simply put into an unusual position of seeing and hearing facts firsthand, bringing with it the responsibility to do my part in awakening the American people to the situation—*as it really is.*

Chapter 3

Shut Down That Pipeline

I have already said that the first time I realized there was no true energy crisis was when the Senator from Colorado visited me in Alaska. However, like many other Americans, I had heard the rumors and hearsay many times before that. In fact, I first became aware of the suppsoed "energy crisis" in 1972 when I was riding on roundup in Wheatland, Wyoming, on a 32,000-acre ranch. That day as we rode in the high country looking for cattle, I noticed a big pump—it was, in fact, a large pipeline that was running across the Rockies. I was curious (that is my nature). I said to the man with me, "Sir, what is that big pipeline running across your property?"

I should explain that because I am a Baptist preacher, I am often called "Brother Lindsey." I suppose it's a courtesy title. My friend answered, "Well, Brother Lindsey, that's one of the major cross-country pipelines carrying crude oil from the West to the East."

"Ah," I answered, "That's rather interesting. I've heard there's a possibility of an energy crisis. I'm sure glad those pumps are running full speed ahead."

That was in 1972. You will remember that 1973 was the first time we were told there was really an energy crisis. The East Coast was used as a test for that energy crisis, and there were long lines of people waiting, *burning fuel while they waited in line for gas they couldn't get.* In 1974, I was again in Wyoming and went to that same ranch. I remember that Fall as we rode roundup over the Rockies, I saw something that startled me. I had just come from the East Coast where I had numerous speaking engagements, and, with the rest of America, I had been told we needed to conserve energy—for if we didn't, we were going to run out of fuel. Crude oil was in low supply and natural gas would soon become a scarce commodity. Imagine my surprise that Fall, as we rode back over that same high country, to find that the big pump was closed down. The pipeline didn't seem to be running.

As we rode the high country on horseback, I asked the gentleman who managed the ranch, "Sir, why isn't that big pump running? You don't mean to tell me that they have closed down a major cross-country pipeline? Back on the East Coast I have seen people standing in line waiting on fuel. What's the story?"

"Well," that old Westerner said, "Brother Lindsey, here a few months ago they came through and started to close down that pipeline, and you know, that thing went right across my property and I believe I had a right to know why they were closing it. After all, I received money from the oil that was flowing through that line across my property, and so I went up to the man and asked him why they were closing down the pipeline. I said to them, 'Don't you know that on the East Coast where that oil is supposed to be going, they have an energy crisis? Don't you know that there are people waiting in line to get fuel and we've got an energy crisis? Man—why are you closing that line down?' "

I listened intently, for I was vaguely wondering if this pointed to some sort of manipulation for a purpose that was unknown to me. The old Westerner went on. "Well, they didn't want to tell me. Brother Lindsey, you know how we Westerners can get sometimes. Cowboys are known for being a little bit mean and ornery, and I decided to use some of that orneriness and persuade that man to tell me why he was closing that pipeline down. So I went up to the boss man and got a little bit rough with him. I told him I wanted to know why that pipe line was being closed down, because after all it was going across my proerty. I let him know that I was an honest American and that I had thought that back on the East Coast they were having an energy crisis, even though we had plenty of fuel out West. Well, the man finally recognized that I was getting a little bit indignant and he said, 'well, mister, if you really want to know the truth, the truth is the Federal government has ordered us to close this pipeline down.' " The old Westerner went on and told how he stood up to the boss man, "Why man, I can hardly believe that. After all, we've got an energy crisis." The boss man answered him, "Sir, we're closing it down because we've been ordered to."

The old Westerner turned in his saddle and he said to me, "That rather startled me. Actually, I had heard there was an energy crisis. It really shook me up. I sure couldn't understand it at all." I confess that I too was shaken. The oil was no longer flowing, and there seemed to be no reason why it should not flow. We were being told that we must conserve energy. The point was being made

very strongly even as we were allowed to
wait in line for fuel.

It is relevant now to go back to the
earlier conversations I had with Mr. X,
who was responsible for developing the
entire East side of the Prudhoe Bay oil
field in Alaska. He was there right
through the entire project, even though
others came in from time to time. He
was an honest man with a fine reputa-
tion, and what was most important to me
was that he was a Christian gentleman.
He did not only say he was a Christian,
but he lived what he said, and he and
I set up quite a friendship. Mr. X was
very definite that the only reason there
was an energy crisis is because one had
been artificially produced.

When I arrived back in Alaska at
Prudhoe Bay in 1974, I said, "Mr. X,
let me relate to you what I saw in
Wheatland, Wyoming, just a few weeks
ago. There was a pipeline going from
West to East across the Rockies, on the
property of a friend of mine. I was riding the range with him in the Fall of
1972 on roundup and the pipeline was
flowing full speed ahead, with all pumps
going. The following year of 1973, in
the Fall, there was supposed to be an
energy crisis, and I found that the pipeline going across the Rockies, one of the
main West-East pipelines, had been

closed down. In 1974, the pumps were not running, .and at that time the man who managed that 32,000-acre ranch told me that the oil companies had told him that they had been ordered to close down that pipeline by the Federal government. Mr. X, if there is as much oil at Prudhoe Bay as in all Saudi Arabia, as you have stated, and if there really is an energy crisis, why was that cross-country pipeline through Wyoming closed down? You must know something about it."

Mr. X. said to me, "Chaplain, I will try to be honest with you today, and I hope it doesn't get any of us in trouble. We are both Christian men, and we can only tell the truth. We, as oil companies, were ordered by the Federal government in 1973 to close down certain cross-country pipelines and to reduce the output of our refineries in certain strategic points of America for the purpose of creating an energy crisis. That really began the first of the control of the American people."

I was astonished at what I was being told. Mr. X showed me the wells and let me know details about the size of the oil pool and the amount of oil that was there. He made the statement that the Prudhoe Bay oil field is one of the richest oil fields on the face of the earth. He said that it could flow *for over 20 years* with natural *artesian pressure,* without even a pump being placed on it. He told me that this was one of the only fields in the world where this is true, and that oil would come out of the ground at 1,600 pounds pressure and at 135°-167°F. He said quite clearly that this was one of the richest oil fields on the face of the earth. He also said that there was enough natural gas, as distinct from oil, to supply the entire United States of America for over 200 years, if that also could be produced.

As I have said, I was astonished. This was during the first year and a half of the Trans-Alaska Oil Pipeline, and the oil companies were supposed to build a natural gas pipeline down the same corridor to supply natural gas to the lower 48 states. The natural gas was to have flowed from Prudhoe Bay to Valdez, been liquefied in Valdez, and transported by tanker to California, Washington, and Oregon, and from there it was to have been distributed across the United States by pipeline.

This was the plan that had been promised the oil companies when they first began the Trans-Alaska Oil Pipeline, and now Mr. X was saying that there was plenty of natural gas here, as well—and as much oil as in all of Saudi Arabia! Yet the media and the Federal government were consistently and continually saying that there was an energy crisis.

I have already shown in Chapter 1 how my eyes were opened. My experience in Wyoming suddenly was seen as part of a widening scope of information. Those experiences in Wyoming—and now my involvement with the Senator and Mr. X—added up to a clear picture of deception and scheming that was hard to understand.

Chapter 4

An Important Visit By A Senator

During the summer of 1975, a Senator visited with me seven days on the pipeline in Alaska. During the three days the Senator was at Prudhoe Bay, I arranged for him to be given a tour of the oil field and facilities. Because of his position in government, he was given an extensive tour. All questions that he asked were readily answered by the oil company executive conducting the tour. The Senator was taken everywhere he requested to go and was shown all data that he asked to see. The Prudhoe Bay oil field, from which crude oil is presently being produced, was explained in detail, and the entire North Slope of Alaska was discussed.

On one of those days we went to one of the drill sites. The Senator asked for more and more technical data and by the time we returned that afternoon to our starting point, we were totally astonished at what we had seen and heard. The Senator had been taken to places that even I as a Chaplain had not previously been allowed to go. However, I stress that I did have executive privileges and could go to any point on the field I wanted to, as well as look at any documents I desired. As I have said, this had been conceded to Chaplains, after about nine months on the Pipeline we were then given executive privileges. We were allowed an executive dormitory and were allowed to see certain things that others could not. Nevertheless, *that* day I was shown things with the Senator and told things by Mr. X that I had not learned before.

We have already explained that the Senator made it clear that the things he had seen that day were in direct opposition to the facts that had been presented by the briefers who came from Washington, D.C. to inform State Senators as to the supposed facts of an energy crisis. I myself was very surprised when I heard the Senator expressing himself, and I said, "Surely a government official would not lie to us about the energy crisis." The Senator answered, "Chaplain Lindsey, we were told something about the Prudhoe Field, and we were told that there was an energy crisis. Today I have found out that there is no energy crisis." It was at that point that he asked me to arrange a further interview with Mr. X the next day, which I did.

When I contacted Mr. X and told him that the Senator would like to talk to him again that day, he said, "By all means. I'll have some time this afternoon, and I'll be glad to give you as much time as you need."

We walked into the office of Mr. X at Atlantic Richfield's facility that afternoon and the Senator began to ask questions. Mr. X was at first a little reluctant to answer the questions, and then the Senator said, "Sir, I want to ask you these questions as a gentleman to a gentleman. I would appreciate very much your direct answers. I promise you that the answers you give will be answers that I would like to use in trying to wake up the American people." Then the Senator went on asking questions. He asked, "Mr. X, what is it that the Federal government is out to do? Why is it that they are not allowing the oil companies to develop the entire North Slope of Alaska? Why is it that private enterprise cannot get this oil out? Mr. X, will you please tell me the whole story?"

What followed included some of the most astonishing answers I have ever heard in my life. This is not opinion, but is actually what I heard from a man who was one of the original developers of the Prudhoe Bay oil field. He said, "Senator, there is no energy crisis! There is an artificially produced energy crisis, and it is for the purpose of controlling the American people. You see, if the government can control energy, they can control industry, they can control an individual, and they can control business. It is well known that everything relates back to crude oil."

The Senator then asked, "Would you please tell me what you yourself think is going to happen?"

Mr. X answered, "Yes, by Federal government imposing regulations, rules, and stipulations, they are going to force us as oil companies to cut back on production, and not to produce the field. Through that they will produce an energy crisis. Over a period of years the intention is that we will fall so far behind in production that we will not have the crude oil here in America, and will be totally dependent on foreign nations for our energy. When those foreign nations cut off our oil, we as Americans will be helpless.

The intention is to create this crisis over a period of time."

The Senator asked, "Mr. X, if you developed the entire North Slope of Alaska as private enterprise what would happen?" Mr. X looked at the Senator and answered simply, "If we as oil companies were allowed to develop the entire North Slope oil field, that is the entire area north of the Brooks Range in Alaska, producing the oil that we already know is there, and if we were allowed to tap the numerous pools of oil that could be tapped (we are tapping only one right now), in five years the United States of America could be totally energy free, and totally independent from the rest of the world as far as energy is concerned. What is more, sir, if we were allowed to develop this entire field as private enterprise, within five years the United States of America could balance payments with every nation on the face of the earth, and again be the great nation which America really should be. We could do that if only private enterprise was allowed to operate freely, without government intervention."

I stress that I am not giving a personal opinion, but I am simply quoting what an expert in the field said.

The Senator was obviously very angry, and he looked back at Mr. X and said,

"Sir, in light of all that you've told me, you've set me thinking today that after being a State Senator for four years, I would like to know something. Sir, will you please tell me what *you* think the American government is out to do?"

It was at that point that Mr. X revealed his opinion that the government was out to declare American Telephone and Telegraph a monopoly, and secondly, to nationalize the oil companies.

The Senator almost gasped at that point and asked, "You mean to tell me that you're convinced that the Federal government is out to nationalize the oil companies?" Mr. X said that was so, in his opinion, and that the Federal government would continue to put such rules and stipulations on the oil companies until fuel prices would go sky high.

That conversation was in 1975. Already Mr. X was predicting over $1.00 a gallon at a time when the American people were reluctantly paying something like 50¢ a gallon. Mr. X told the Senator and me that the Federal government would force oil prices to over $1.00 a gallon, and in doing so would make the oil companies look like villains, and the American people would request the Federal government to nationalize the oil companies.

Mr. X gave facts and statistics that

day, and in the last six months of the construction of the Trans-Alaska Oil Pipeline, it became clear that he certainly knew what he was talking about.

The Senator had another question. "Mr. X, if you're convinced that the Federal government is out to nationalize the oil companies, undoubtedly you have a target date?"

Mr. X said, "Yes, Senator, we do. As oil companies we have already calculated that with present government controls and regulations, we as oil companies can remain solvent until 1982." Those were Mr. X's exact words.

The Senator said, "Sir, I'm amazed at what I've heard, because it falls in line with what I've believed for years, in what the Federal government and its agencies are really attempting to do to the American people."

The Senator was obviously very upset, and as he discussed it all with me in the dormitory room later that day, he said that when he went to the lower 48 states he would attempt to have somebody publish the truth of this matter and use it in their election campaign. He wrote a personal letter to Ronald Reagan and received a personal reply—the Senator wanted Ronald Reagan to go to the North Slope of Alaska and see the truth as he had seen it, and make the energy crisis a major platform in his campaign. He believed that if he did so, he would be elected.

Ronald Reagan wrote back to the Senator and said, "Sir, I'd like to, but I don't have. the time—my schedule will not permit." The Senator attempted to get others to know the truth about Prudhoe Bay oil field and the fact that there was no true energy crisis, while something could still be done before the *created* crisis became even more severe. It was artificially produced, of course, but many of the American people were becoming convinced that there really was an oil crisis, while the oil companies themselves were constantly being hamstrung.

The Senator could not get anyone willing to stick their necks out far enough to tell the truth because this was becoming a major issue. The American people were being affected, gasoline tanks were empty, crude oil was in short supply, and even natural gas in certan of our East Coast cities was cut back that year to such a low level that homes were going cold. By creating an artificially induced energy crisis, the American people in large numbers became convinced that our energy really was short.

In our last chapter, we told about that pipeline in Wyoming. The oil was available, but the pipe was shut down. As we proceed, we shall see that huge quantities of oil were available in Alaska, and could readily be made available to the outside world, provided the pipeline itself was available. We shall see that intensive efforts were made to hinder that work— to slow it down, to increase its costs, and all the time to hoodwink the American people.

What was behind it all? It is not enough simply to say that the current President is at fault. These regulations were proceeding before he was President, indeed, during the term of a President who represented another Party. This scandal I am exposing is something that leads to the bureaucratic controls behind —and yet beyond—government political leaders, as such. I shall have more to say about that as we proceed . . . and about important financial operations.

What was the involvement of the New York banker and of those Arab Sheiks who had to help bail out the oil companies when they faced bankruptcy? These are questions to which we must have answers. At the appropriate point we shall give you more of the facts, but first we turn aside to give you some information about the oil fields themselves and how they work, and then (in Chapter 7) give some typical examples of the

wasteful expenditures forced on the oil
companies. These examples could be
multiplied.

We shall refer to the problems with
the Unions, but those were relatively minor. The oil companies could have lived
with those frustrations, but we shall still
give an illustration of that problem area,
so that the whole picture is brought into
clearer focus. Then we shall go on to
the far greater problems involving the
ecology.

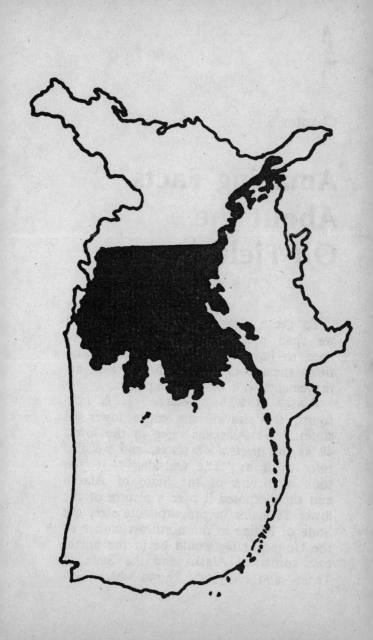

Chapter 5

Amazing Facts About the Oil Fields

To get a clear understanding of what we shall present in later chapters, we need to have a clear picture of the oil fields themselves and of the working arrangements with the oil companies.

Alaska is a huge state. It is one-fourth the size of the entire lower 48 states. We Alaskans refer to the lower 48 as the *original* 48 states, and we also refer to it as "The Outside." If you took a picture of the State of Alaska and superimposed it over a picture of the lower 48 states in proportionate size, the State of Maine in the northeast corner of the United States would be in the northeast corner of Alaska and the State of Texas—and everybody knows where Texas is (just ask a Texan!)—would be on the southeastern coast of the State of Alaska. Alaska is the largest state in the United States, yet 60% of the population of Alaska is in the one city of Anchorage.

Alaska has three major mountain ranges; the Rockies, the Kuskokwin, and the Brooks Mountains. As you travel northward over each mountain range, there is a climatic change. The southeastern coast of Alaska is known as the Osh Kosh, and this area of Alaska is very mild in winter. The Japanese current which warms Washington and Oregon also keeps this area of Alaska mild. Immediately after crossing the Rocky Mountains into the first interior area of Alaska the winters become severe, going to 50° and 60° below zero. After crossing the second mountain range you come to the Arctic Circle area. The Arctic Circle is an imaginary line around the face of the earth, north of which there is at least one day per year when you have 24 hours of sunlight and another day when the sun never appears above the horizon.

Just north of the Arctic Circle are the Brooks Mountains, and north of the Brooks Mountains is the area to which we are referring in this book as the North Slope of Alaska. This North Slope is a vast Arctic plain, many hundreds of

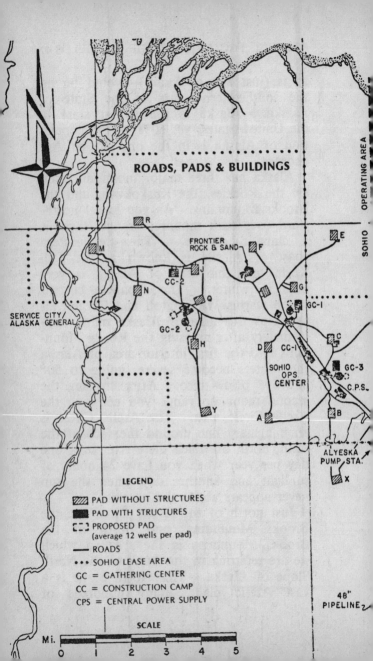

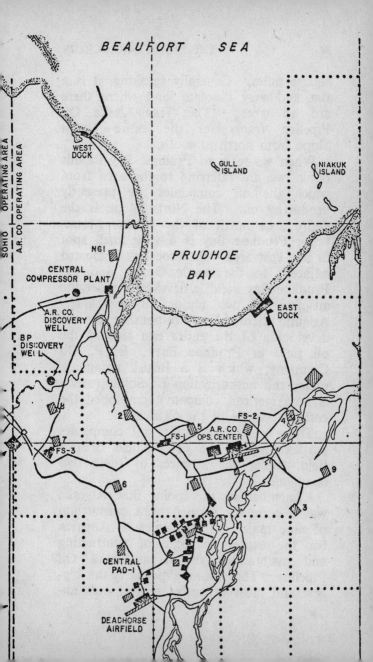

square miles.   Generally speaking, it is a
flat and very desolate land where there
are no trees.   The Trans-Alaska Oil
Pipeline transverses the entire North
Slope from north to south.

When we refer to Prudhoe Bay in this
book, we are referring to the area from
which the oil companies are presently
producing oil.   The North Slope is the
entire area north of the Brooks Mountains; Prudhoe Bay is a very small spot
in this vast area.   Prudhoe Bay is located
adjacent to the Arctic Ocean, and the
Prudhoe Bay Field is developed under the
auspices of two major oil companies.
Atlantic Richfield was responsible for the
developing of the entire east side of the
oil field at Prudhoe Bay.   B. P. Oil
Company, which is a British company,
under the authorization of Sohio (which
is an American company), developed the
entire west side of the oil field.

There were seven other oil companies
participating in the development of this
field, under the auspices of these two
companies.

Remember at this point that Alyeska
was a company formed by a consortium
of nine major oil companies of America
for the express purpose of contructing
and maintaining the Trans-Alaska Oil
Pipeline.   The Alaska Pipeline is the biggest and most expensive project ever un-

dertaken by private enterprise in the history of the world.

When the oil companies began to develop the pipeline route north of the Brooks Mountains, there were no people, no roads, and no towns. There was nothing but a vast Arctic wilderness. This is especially relevant to the problems forced on the oil companies by the Federal and State officials in regard to the whole matter of ecology and environmental protection.

At tremendous cost to the oil companies, entire self-contained cities were flown in by Hercules aircraft and then constructed to house three to five thousand workers each. As there were no people, no roads, and no air strips, the huge Hercules aircraft landed on frozen lakes in the winter time. The equipment was assembled, gravel pads were built, and the housing units and all life support systems were constructed on the gravel pads. Everything was brought together right there—all electrical systems, water systems, sewage systems—everything had to be constructed on the actual sites.

Hercules aircraft are huge four-engine turbo-prop aircraft, capable of carrying tremendous loads. The entire rear section of the aircraft opens and very large objects can be placed inside. In fact, the Hercules was designed by the military during the last World War for the purpose of driving tanks and other military craft directly on board. Again, as we proceed, we shall see that these huge aircraft were at times used in ways that can be best described as frivolous, adding huge costs to the oil company expenses, and ultimately adding to the price that you and I, the consumers, will be paying at the gas tank.

In 1974, the cost to the oil companies of one Hercules was $1,200 per hour to rent. Remember, not one penny of government money was used for construction of the Trans-Alaska Oil Pipeline. It was entirely financed by private enterprise.

Animals north of the Brooks Mountains on the pipeline corridor had never seen human beings. The caribou, bear, and Arctic wolves had never seen man and had no fear of man. Almost every day you would see a survey team sitting in one of the few trees while a bear went by.

North of the Brooks Mountains the ground is known as permafrost, because it is perpetually frozen all year round. In the area of Prudhoe Bay the ground is frozen for 1,900 to 2,100 feet down from the surface. Yet to the depth at which the oil is produced, which is approximately 8,700 feet, the oil will come out of the ground at 135°F.

Most oil fields in the lower 48 states have to be *pumped* from the time of their original production, and we are often told that this is a major reason why America imports oil from such places as Saudi Arabia. The argument is that because the Arabian oil is so readily available and so much easier to bring to the surface, it is ultimately less expensive to import the oil than to take it from our own ground. However, that is not the case at Prudhoe Bay; indeed it is not the case on the entire North Slope of Alaska. After 20 years of production at *natural artesian pressure*, the oil companies will inject treated water into the pool of oil, and then they can continue production at natural artesian pressure for many years to come.

One of the leading news magazines recently stated that the Prudhoe Bay Oil Field will run out of oil within five years. This is totally contrary to all technical data that I saw. In fact, the Prudhoe Bay Field will produce well over 20 years without any artificial methods, and then for many years to come at a rate of approximately 2 million barrels of oil every 24 hours. We stress that this is oil available from only one pool of oil; keep in mind that there are many, many proven pools of oil on the North Slope of Alas-

ka. At the end of this book we will tell
of one field that has already been drilled
into, tested, and proven. Yet the Federal
government ordered that no oil be produced from that new-found field. We
shall elaborate on this in detail later.

There is an interesting point to mention
in passing. Though the ground is frozen for 1,900 feet down from the surface
at Prudhoe Bay, everywhere the oil companies drilled around this area they discovered an ancient tropical forest. It was
in frozen state, not in petrified state. It
is between 1,100 and 1,700 feet down.
There are palm trees, pine trees, and
tropical foliage in great profusion. In
fact, they found them lapped all over
each other, just as though they had fallen
in that position.

What great catastrophe caused this
massive upheaval, and then led to such
dramatic changes in the climate? We
stress again that everything is frozen—not
petrified—and that the whole area has
never once thawed since that great catastrophe took place. So what could possibly cause these dramatic happenings?
Most Bible scholars would come to one
of two conclusions. Some would argue
that it is tied in some way to a great ice
age which they believe occurred between
Genesis 1:1 and 1:2, when many events
took place that are not thoroughly under-

stood. Others would point to the catastrophic effects (and aftereffects) of the Biblical flood of Noah as the case, suggesting that this is evidence of a sudden overtaking by the flood waters and sediments. The breaking up of a great canopy of water that once surrounded the earth, as well as the breaking up of the great "fountains of the deep" referred to in Genesis, could easily account for the tremendous volume of water that since then encompasses the globe. It is believed that the resulting atmospheric and geologic changes were the cause of the drastic changes in climate.

It is interesting to notice that tropical ferns have also been found at the Antarctic, and the evidence from these two areas, considered together, certainly suggests that there has been a dramatic change from a worldwide tropical climate to an Arctic climate within datable times.

It is also interesting to remember that the great Arctic explorer, Admiral Byrd, reported seeing tropical growth in near-Arctic regions. Most write this off as being some sort of a mirage, or maybe even an hallucination, but perhaps we have to reconsider. Just as there can be a beautiful grand oasis in the middle of the desert of Egypt (such as the Fayum Region), perhaps there have been oases in this other kind of vast expanse in the

Arctic Ocean area, where these subterranean tropical plants are (for some as yet unknown reason) still growing on the surface.

The finding of underground tropical growth is not hearsay, for I have personally watched these palm trees and other types of tropical plants being brought to the surface. Let me give you two examples. One day I watched as a pine cone was brought up from a well (although not considered tropical, they apparently grew together in historic times), and when we first saw it, it looked just exactly as it would look on a young pine tree today. It was closed, and we put it in an office on the premises of Atlantic Richfield. We simply put it on the desk and left it. The next day we came back and the pine cone had opened up. You could quite clearly see the seeds on the inside of the cone. This was obviously after thousands of years of being in a frozen state, hundreds of feet beneath the surface.

I personally have palm fronds in my home which were brought up from some 1,700 feet below the surface. Again I would like to make an observation, without necessarily giving an opinion, because I do not regard myself as expert in this area. I simply want to state that consistently this tropical forest was found between 1,100 and 1,700 feet beneath the surface. The actual base of the *perpetually frozen* ground is approximately 200 feet below the depth of the frozen tropical forest. The oil is found at a depth of 8,700 feet, average, and it is amazing to realize that it comes from that depth *without artificial pumping.*

I want to tell you a second incident that you will find hard to believe. As it cannot be documented, it might not be true, but I shall simply report it as it was told to me. One day I actually watched an operation proceeding at Pump Station 3, but did not take any special interest. After all, proceedings were going on all the time. However, on this particular day a man whom I personally know to be very reliable came to me and said something like this: "Chaplain, you won't believe this, but we were digging in this gravel pit on the Sag River, quite a number of feet under the surface depth. We brought to the surface what looked like a big Louisiana bull frog. We brought it into the building and allowed it to thaw out."

As I say, what was then told to me is hard to believe. However, let me point out that the frog is a cold-blooded mammal, and that in the winter season it *does* go into a virtual state of deep freeze— much like the hibernation associated with bears and other Arctic animals.

This man described the way in which the frog was left there and then thawed out. He claimed they actually watched as it totally thawed, and that it then quite perceptibly moved—in fact it appeared to be alive, with those perceptible movements taking place for several minutes. Then the movement ceased, and the men threw the frog away. Of course, it would have been better if they had kept it and had the story both witnessed and properly authenticated. Nevertheless, I mention it as an incident that was accepted by others as actually taking place. I have no reason to doubt it.

This then is the setting for the North Slope of Alaska. It is a land of extremes, and that is well-illustrated by its temperature. At Prudhoe Bay I have seen it go, with the chill factor, as low as -130°F (130 degrees below zero). I have also seen it go higher than 90°F in the summertime (this being above zero and quite hot, of course). It is a beautiful land—a land that I have learned to love. In fact, during the months of July and August, the area of Prudhoe Bay is one of the most fabulously beautiful areas of the world. It looks like one great vast golf course, stretching for hundreds and hundreds of miles.

Anyone for golf?

Chapter 6

The Workings of An Oil Field

We have said that the Alyeska Pipeline Service Company was a consortium of nine major U.S. oil companies. Each of these sent a certain number of their executives to Alyeska for the contruction phase of the Pipeline. This meant that we had men from each of the nine oil companies who had been placed in management positions spread all across the North Slope of Alaska. These men would work so many weeks on the job, then work a number of weeks back home —and then they would return to the job in Alaska again. This meant that there was a continual rotation of executive officers, and, in practice, it was a very effective system. A man was not subjected to the rigors of the Arctic all the time, but would come back refreshed and able to perform with top efficiency while his alternate was relaxing in the lower 48 states or at Anchorage.

Most of the relaxing was done at Anchorage, rather than taking the arduous trip to the south at very regular intervals. It is relevant to point out that the top executives in the oil company worked one week on and one week off in rotation. The further down the ladder you went, the longer they stayed on the job and the less time they had at home. By the time you got to the ordinary worker on the Pipeline, he was expected to stay on the job for six or seven weeks at a time, to go home for one week, and then to come back for a further six weeks.

The top executives would always overlap each other for one day, so that there was constant briefing and debriefing. It was thereby insured that the work would proceed without undue problems. It was at these briefings that I constantly gained a great deal of information. I spent a lot of time in the offices, and at no time did the executives object to the fact that I was present when they were talking about activities that were proceeding at that particular time. It was not my goal or purpose to be there to "gain information," and indeed if I had been there for that purpose, I would have taken very much more notice and kept much more elaborate records. At that time I did not even realize just how pertinent the information really was.

Neither did I ever think that *our own* Federal government would go this far in producing an energy crisis. As the Pipeline was nearing completion, I then personally realized just how critical all this information really was. The total picture did not fit together until the end, and in fact it has not yet all fitted together. I confess that there are aspects that I simply cannot rationalize. I do not profess to have all the answers. This is one of the reasons why I have deliberately set out to report first what I know to be fact, before I briefly set forth my own opinions or speculations. Of one thing I *am* convinced. Somewhere, some place, there definitely appears to be a conspiracy.

Because there were, of course, numerous high officials, and each of these was rotating with his alternate, obviously a great deal of discussion took place. Statistics and figures were thrown around like confetti, and some of it landed on my shoulders. Perhaps we should change that and suggest it was thrown around like a basketball. Sometimes the ball landed in my lap, and I took it and ran with it.

Despite the implementation of rules and regulations in ways that were unbelievable, the major development of the Pipeline took place so rapidly that at times information was available which was quickly withdrawn. One outstanding example of that was the whole matter of Gull Island, of which we shall give full details in a later chapter. We shall see that the information relating to Gull Island was ordered to be sealed by the government authorities within days after proof of the find.

It is not our purpose to give all sorts of details as to the day by day administration of the Pipeline, or of the human nature of the men. There were, of course, the common problems such as theft, with the usual attitude of, "You scratch my back and I'll scratch yours." That is in all big business and government operations, wherever human beings are found working—around the face of the globe. Human nature does not easily change, whether those concerned are in Alaska or in the lower 48 states.

The sort of graft that so often is associated with private enterprise and big companies is prevalent in many areas. In fact, ultimately human ambition demonstrates itself in ways that have similar roots, if only we can get back and understand the scheming *behind* various operations. Some people are anxious for financial gain; others are more interested in a power structure; and when it comes to the political arena, that power structure might go way beyond mere money. It is possible to relate this to the oil fields, and to see some semblance of comparison with what is taking place in Canada.

Canada has already nationalized its oil companies. That is an actual fact of history, and this was often referred to by executives of the oil companies working for the Pipeline. Often I heard it related that the same patterns that were used by Canada for the nationalization of their oil companies, appeared to be the pattern that the United States government was following in its dealings with oil companies today. The oil company officials in the top echelons have suggested that the Federal government wishes to nationalize the oil companies of America. We will elaborate on this in detail in a later chapter of this book.

The heading of this chapter is "The Workings of an Oil Field." It is relevant to emphasize that the United States government, as such, did not own anything —equipment, machinery, buildings, or anything else—on the oil fields. Not one penny of government money was invested in the Pipeline, yet the government exerted all sorts of pressures as they implemented their multitudinous rules and regulations. Neither did the oil companies own all of the equipment, for in many cases the work was subcontracted, and

often the machinery was owned by the
company to whom the work was contracted.

One official was responsible for all the
subcontracting of heavy machinery on the
east side of the oil field. At one point I
heard him state that in a 30-day period
he gave out as much as $2 million dollars
in contracts for lease of equipment.
That man's work is uniquely different
from anything else, anywhere on the face
of the globe, and that is true of so many
jobs associated with the oil fields on the
North Slope of Alaska. Because of the
Arctic climate, many positions have been
created and developed that have no paral-
lel at all in any other project. Very often there is no available training, such as
with university degrees, for the job requirements are unique to the Alaska oil
fields, and there certainly is no university
found out in the tundra on the North
Slope!

I know of one man who was a sheep
herder in Wyoming, and he operated a
huge ranch. He came to Alaska because
he heard of the exorbitant wages on the
Pipeline, and he wanted a slice of the
cake. He started as a general worker at
the very inception of the Prudhoe Bay
Oil Field, and today he is an invaluable
executive with Atlantic Richfield (ARCO).
He had no specific training—he was

trained on the field, and I personally heard him say that he cannot be transferred because there is no other job like his at any other place on earth. This man is so unique that he virtually knows where every nut and bolt is at Prudhoe Bay, and he is quite irreplaceable. Mr. X remarked to me one day that if he ever wanted anything, he would simply go to this particular man. He seemed to always know where everything was.

Such a man *is* invaluable, if only because of the high turnover of the labor force on the Alaska North Slope. Many of those who had been there for comparatively short periods of time had no idea as to what had gone on before they had arrived, or as to the way certain activities developed. Over and over again the very nature of the field demands training that is simply not available anywhere else. This can be found only in the "University of Hard Knocks." The Alaska oil fields certainly is one big branch in *that* University!

The dorms in which the men lived in the camps were very well appointed. There were two men in each room in a 52-room section. Men shared common baths in these common dormitory areas. As that executive stated, the food was the best you would find anywhere in the world. During the first year of pipeline

construction, it was not unusual to have
steak and lobster twice a week. I sat
one evening and watched a man eat two
steaks, and then he put one in his lunch
sack so that he would be able to carry
it off to eat on the job the next day.
Nowhere but on the Trans-Alaska Oil
Pipeline would you see a welder heating
up his steak out on the job with a welding torch, while the steak was on a big
piece of metal. He was actually heating
the steak from the bottom side of the
metal!

The food was always in plentiful supply, being available 24 hours a day, seven
days a week. The men did not pay for
their food, nor did they pay for their
candy bars or pop. They simply took all
they wanted.

Another thing the general public does
not know is that everything the men
earned (after taxes and deductions) they
could take home with them, because their
dorms and all food were free. It was
not unusual to see a weekly take-home
paycheck of $1,000 after all taxes and
deductions had been taken from the salary. In fact, the largest paycheck I saw
for seven days of work was actually over
$3,000 for an ordinary working man.
Workers on the oil field did not exactly
starve—in fact, most people would consider that their conditions were very desirable.

Chapter 7

Toilet Paper Holder for Sale Cheap— Only $375.00!

We said we would mention problems. We do not wish to major on Union difficulties, so we shall give only one example to keep the picture in true perspective.

I was sitting with Alyeska's field engineer in the office, simply shooting the breeze before getting down to more important business. In walks one of the workers and says, "The toilet paper holder is falling off the wall in the commode stall over yonder in B dorm."

"Okay," said the manager, and he called in a carpenter. The carpenter came in, dressed for work, of course.

"Hey Jim, I'd like you to go over and fix the toilet paper holder in B dorm." "Okay," said Jim and off he went. I watched him go out and vaguely thought that he looked a capable man, really dressed for the part. I thought of some of the carpentry jobs around my home I'd like him to do. Surely he would be a lot quicker than I would be, although before very long my opinion on that was drastically changed.

The manager and I went on discussing our business, and had forgotten about that unimportant toilet paper holder over in B dorm. Forty-five minutes went by, and Jim, the carpenter, returned.

"Hey," he said, "I can't do that job over there. That's a metal wall and it has to have a screw put in it. That's not a carpenter's job—you ought to know that. That's a metal worker's job. The union would not let me do that."

You notice that it had taken him 45 minutes to decide that, and he then came back to the office. Of course, we must allow the man to have time off for coffee and a cigarette. However, I did think 45 minutes was just a little long.

"All right," said the manager, and he did the expected thing and called over a metal worker. In due time the metal worker arrived, and he in turn was told of the urgent need to repair the toilet paper holder on the metal wall in the dorm. Off went the metal worker, and about an hour later he came back. I was still there, for there were some matters that I needed to go over in detail with the manager. In walked the metal worker, and now I had a job to control myself.

"Hey, I can't do this. This involves a screwdriver. That's a laborer's job, and I'm a metal worker. I just tie metal together. You can't expect me to do a laborer's work."

The manager was beginning to feel frustrated, though not all that much, for after all these things happen so often. "All right," he said, "I'll send for one of the laborers." And he did. A little while later a laborer came in, and the manager carefully explained to him what dormitory it was he was to go to. He was very particular, because he had the impression that the man might not be following him very closely. The laborer went off, apparently knowing what it was all about, and the manager and I got down to our business again. It was probably 40 minutes later that again we were interrupted, this time by the laborer coming in with his story as to why he could not fix that toilet paper holder on that metal wall in B dorm.

"Hey, you can't expect me to do this.

That screw you talked about—that's gotta
go into some wood there—you know that
as well as I do. That's a carpenter's job
—I'd be on strike if I were to go a-
gainst the union rules in a thing like
this."

The manager turned to me, this time
really frustrated. "What do you do,
Chaplain? The carpenter can't do it because metal is involved, the metal worker
can't do it because there's a screw involved, the laborer can't do it because
there's a piece of wood involved—what
do I do with that line up of men who
are wanting to use the toilet paper?"

In desperation the manager now called
in the foreman of the metal workers, the
foreman of the carpenters, and the foreman of the laborers, hoping to be able to
figure out some way in which somebody,
somewhere, somehow could fix that toilet
paper holder onto the metal wall with the
little bit of wood over in B dorm.

So, these foremen came in, each of
them being paid about $25.00 an hour.
The carpenter would have earned something like $15.00 an hour, the metal
worker about the same, and the laborer a
little less. So the foremen were called in.
The doors were closed. Chairs were
drawn up. They sat down to this very
important conference. None dare inter-
rupt. It was almost as though the blinds

should be drawn in case anybody would
happen to see over their shoulders as
they seriously discussed regulations for
putting toilet paper holders on walls—no,
not just walls, metal walls with wood
protruding.

At last an amicable arrangement was
entered into. It was clearly an excellent
illustration of the unity that could be
shown by human beings when they set
their mind to do a thing. Nothing is too
hard for men to accomplish when they
really are serious about finding a solu-
tion! The conference relating to the toilet paper holder was a glorious demon-
stration of human ingenuity, friendship,
and common sense. (Or was it?)

Of course, you will be very interested
to know what the result was. When we
tell you, it will be something like the in-
terpretation of the Pharoah's dream in
the days of Joseph. Once the interpreta-
tion is given, it is obvious.

The decision was that the foremen
would call up one man from each of
their ranks, and those three men would
go together to that metal wall with the
wood protruding over in B dorm. There
was no decision made as to who would
actually lift the toilet paper container, but
it was agreed that the three foremen themselves would be there to insure that nobody did anything that was against the

union rules. So the procession went across to B dorm. Unfortunately, the manager and I were unable to go . . . *we couldn't stop laughing long enough!* To be honest, we found it hard enough to not laugh until the team of valiant workmen were out of sight. Then we laughed until they came back.

We were told later what happened. One man would pick up the screwdriver. The other would pick up the piece of wood. The other would hold the screw. Between them they eventually managed to get the toilet paper holder back onto that metal wall with the piece of wood protruding, without offending any union rules. The three officials were satisfied, the workmen were pleased with their noble day's work, and the line of men that had congregated at the other toilets was reduced as the word went around that the toilet in B dorm was again in working condition.

As we say, everything can be done so long as there is a spirit of compromise, fraternity, and "ridiculosity."

You think that's the end of the story? Well, it's not, actually. After all, rules are rules. History has that grim habit of repeating itself. Who knows, perhaps one of those three men did not do his work properly. It would be a dreadful thing to go into that room and find the toilet paper holder had fallen off again. Perhaps by that time one of the foremen would be gone, and they would not have a proper reference to be able to see the matter through so expeditiously and so harmoniously as it had been the first time.

The manager was a man of great foresight. He recognized the problem, and so he said to the men concerned, "Now that you men have done such a good job, and have come to such a wise conclusion, we must see that this is properly established in case there's a repeat at some future time. I must put this down and telex it for our records." He did just that, and sent an elaborate telex down to Fairbanks. Presumably someone at Fairbanks had the arduous task of deciding into what subsection the new regulation should be inserted in the New Operations Manual.

Looking back, it is undoubtedly funny, and I've laughed many times as I've thought of that particular incident. However, the more serious aspect is that the cost of replacing that toilet paper holder on that metal wall with a small piece of wood attached was astronomical! (And that didn't even include the cost of buying the holder, itself.) I have actually sat down and calculated what the total cost would be, based on the salaries of the men concerned. Six men were involved, at salaries ranging between approximately $12 and $25.00 per hour, so the total cost was something like $375.00. As we say, it has its funny side, but it was a ridiculous, frustrating waste. Unfortunately, that was typical of so much that took place on the oil fields.

By the way, the next time you go to a gas station and pay over $1.50 for a gallon, *remember that toilet paper holder.* Your extra cents are helping to pay for that important piece of engineering, and that is symptomatic of so much that took place while the Pipeline was being constructed.

As we have said, there were many problems over union matters—as with various types of labor being required for the simple maintenance of vehicles. There were many irritating delays and unnecessary, exorbitant costs.

The practice of wobbling became a serious problem. That was what the union men called it. It seemed that everything was piling up, all at once. It seemed almost as though there was some underlying force planning this whole thing—every day another catastrophe. By now there were only six months to go until the flow of oil, but everything was breaking loose —the whole place was coming apart. The unions had agreements with the oil companies, and they had promised that for the life of the pipeline they would not strike. The reason that they had promised this was that the men had been given salaries that were exorbitant. Nowhere on the face of the earth could you make that kind of money in these trades, and therefore the unions agreed to sign an agreement that they would not strike.

And then, some of my own Christian men—men who were supposed to be honest—came to me and said, "Chaplain, we can't strike, but we can wobble."

I asked, "What's wobbling?"

They said, "That's just another way of striking. Instead of leaving the job and not getting paid for it, we just slow it down. We just sit in the buses and refuse to work because conditions are not right."

Who told them that the conditions were not right? Those conditions had been right for two years, and in all that time there had been no wobbling. The conditions were identically the same as they had been through that period of time, so who was telling them that conditions were not right? Why did they decide to start wobbling?

When I asked for further explanation of this term "wobble," they said to me. "Haven't you ever seen a wheel turning on its axle? It doesn't come off, but just wobbles and slows the whole thing down." I said to myself, "That's it. That's exactly what's happening. They're trying to slow the whole thing down."

So Union problems were adding to other problems, such as the demand to dig up the pipes, the constant urging for withdrawal of permits, the claims that there were faulty welds, and the attempt everywhere to stop the flow of oil.

Despite these problems, it is worth mentioning that to a great extent the lower echelons of workers were very much behind the oil companies, especially in these last 6 to 9 months. They recognized that the government policies were ridiculous, and they could see what was happening. It was talked about quite openly. However, those workmen did not have the in-depth understanding I had, for they did not have executive privileges which I had as Chaplain. It is true to say, however, that to a remarkable extent the workmen were very upset at the ridiculous impositions by government authorities.

It is also true to say that the government policy was to put restrictions in the way of the oil companies at every conceivable and every inconceivable point. They seemed determined to give problems everywhere they could. It was bureaucracy gone made.

The oil companies put some information out from time to time in their periodicals, but their reports are not usually available to the general public, and although much of the information about the way the ecology was protected to such extremes was written up, it did not receive wide publicity.

Extremes? Yes—let us illustrate that.

Chapter 8

Want Some Falcons? Just Two Million Dollars ...A Pair!

The manager at Happy Valley Camp called me into his office one day (by the way, his name was Charlie Brown, and I always did like Peanuts!) By this time I had begun to notice that *some* things simply didn't make a lot of sense. Costs seemed to be exorbitantly high, and as time went by I was to find that this was indeed true in all sorts of strange ways.

The initial constructions phase of the Trans-Alaska Oil Pipeline involved building a road from Fairbanks, Alaska, to the Arctic Ocean at Prudhoe Bay. This road is approximately 400 miles long. It is a gravel two lane road, right on top

of the tundra. On this Northern Sector
of the Pipeline there were no roads, no
people, and no towns. Alyeska Pipeline
Service Company had to construct everything from scratch. This road from Fairbanks to Prudhoe Bay is commonly referred to as the haul road.

On this Spring day the haul road was
being constructed across a certain area.
It is important to know, so that this
story will be understood, that the Trans
Alaska Oil Pipeline haul road that ran
from Fairbanks to Prudhoe Bay was so
designed that it would affect the ecology
as little as possible.

This might seem strange to most people
in the lower 48, that is to say, all the
states excluding Alaska and Hawaii, but I
have actually seen a 'dozer driver lose
his job just because he accidentally drove
the 'dozer off the main path of the road
and drop out onto the tundra. That's
how particular the ecology people were
about the protection of the precious tundra. We shall discuss the ecology and
environmental protection a little later, but
at the moment let us simply say that in
the construction of the Pipeline there
were many ecologists checking on everything. There were Federal government
men, as well as State men, and sometimes you would find these men actually
walking out in front of equipment so

that they could move away little ground squirrels to make sure that no animal was affected in any way by the building of the haul road.

So this day I was called into Charlie Brown's office at Happy Valley Camp, and he said, "Chaplain, you've just got to see what's going on here. I just wasted two million dollars."

I looked at him, wondering what he meant. He did not seem to be too unhappy personally, and I knew that he was talking about the company's money and not his own.

"Never mind," I joked with him, "With all the money you've got, you won't even miss a couple of million. I must come to you for a loan myself sometime."

The manager smiled, but then he became more serious. "Chaplain," he said, "We talk a lot about the way this Trans Alaska Oil Pipeline cost overrun is getting out of hand. I told you that originally the Pipeline was supposed to cost $2 billion dollars, and that the cost overrun is building up every day. Well, sir, as you know, we are putting this haul road across the hillside just outside Happy Valley, and we've been given permission by the government to build the road there. It's not as though we didn't have permission—we've gone through all the right channels, and we're putting that haul road across that hillside, and we have no reason to doubt that we could get the project done in good time."

He paused, and I wondered what was coming. I looked up and saw that he seemed really angry about something.

"What's bothering you, Charlie?" I asked him sympathetically.

"Well, you'll never believe it. There was a falcon's nest up on the top of that hill. You know as well as I do that the major nesting grounds of the falcons are the Franklin Bluffs and around this Happy Valley area. These ecologist creeps want to insist that the falcons along the Sag River are on the semi-extinct list, and that they can't be disturbed at any cost. Now we find there are those two falcons nesting up there. One of the (- - - -) ecologists found them, and he told us we'd have to stop the whole job."

"The whole job?—you're not serious!" I asked.

"Never more serious in my life. This creep found them, and he told us we had to stop the whole job—I mean he told us we'd have to shut down everything, with all those hundreds of men out there on the job working. That guy had the authority to tell us we couldn't go on with our construction, even though we'd been given permits to build it this way, and we were deeply involved with hundreds of men at work.

" 'Don't give me that nonsense' I said to him. 'You don't really think we're gonna' stop all this work just so a falcon can sit on its eggs?'

" 'That's exactly what I am saying,' he said. This creep told me, 'You can't go on with this construction until the falcons have finished nesting.' "

"Why can't you move the (- - - -) falcon's nest further across the mountain?" Charlie asked him. That seemed to me to be a sensible enough question.

"My job is to protect the falcons. I'll do my job, you do yours. The road doesn't go through until those falcons have finished nesting." Charlie was told.

Charlie Brown looked at me, and obviously he didn't know whether to laugh or to cry. "Can you really believe it? What could I do? He's got that big book of rules and regulations, and if I go against him not only do I lose my job, but the company gets fined, and the road doesn't go through anyway. They have got all these rules and regulations, and the overrun is simply getting to a stage of being absurd. This is the greatest construction by man in all the history of the world—so the experts tell us—and yet some creep can tell us that we can't build our road until two falcons have finished nesting!"

"So what did you do, Charlie? Did you punch him in the nose?" I asked, with a rather un-Chaplain-like suggestion.

"No, that wouldn't have done any good. He's got both the Feds and the State on his side. I don't have any choice. I had to apply for another permit and reroute the whole (- - - -) road. We couldn't wait a month for the falcons to get through with their breeding process, so we just had no option but to reroute the whole haul road. Chaplain, we had to reroute the whole road all the way around that hill, and around the other hills, and take it away from Sag River, and then haul the gravel that much further."

I looked at Charlie Brown, and despite the seriousness of the situation, I saw the funny side and I laughed. "Sorry, Charlie, but it's so ridiculous I can't help laughing." I wiped the smile off my face and then I said more seriously, "How much do you reckon it will cost to move around those two falcons?"

"Well, I've actually calculated it. In order to go around that one nest, it's going to cost the oil companies an additional $2 million dollars. What do you think of that?"

I said to Charlie Brown, "Sir, wait a minute—are you telling me that because of those two falcons the oil company is going to be charged an extra $2 million dollars—$2 million dollars *extra* for the cost of that road—*a million dollars a falcon?*"

Charlie Brown nodded his head and said, "Yes, that's correct. Two million dollars—a million dollars for each falcon."

I could hardly believe what he said as it sank in. I said to him, "Do you think they'll ever come back to this particular spot—are they likely to come back there to that nest?"

"No," he said. "Nevertheless, we can't wait a month, and those creeps wouldn't let us move the falcons. After all, Chaplain, that would be a national crisis, and we must salute the flag and all that, you know. So we'll just quietly have to put up with it. Of course, when you go to fill up your car with gas, remember those two falcons—*you're* going to pay those extra $2 million dollars that we had to spend to reroute the haul road to protect the two falcons on the hillside outside Happy Valley. Maybe it won't be just you, Chaplain, but you and your friends will pay that $2 million dollars."

I love animals and living things, and I think they should be protected, but I do think that these things can be taken to a ridiculous extreme.

Some time after this I was in the lower
48, in the middle of a series of speaking
engagements across America each winter.
On this occasion I stopped off in Seattle
to stay with some relatives of my wife,
and we were sitting at the breakfast table
one morning with the radio on. I heard
an editorial. I think it was three minutes
long, if I remember correctly, and it was
by the Sierra Club.

By this time I had been to Prudhoe Bay
for one winter and two summers—a year
and a half. I had seen the caribou mi-
gration, I had watched the geese and the
ducks come to the North Slope by the
thousands. I had seen the beauty of the
tundra in the summertime, I had watched
the fantastic specter of the Northern
Lights, and I had enjoyed the snow in
the wintertime—in fact, I love Alaska,
because I'm a natural born outdoorsman.

I had been very interested in all the
ecology measures the oil companies were
taking to protect the North Slope while
they were building the Pipeline. I had,
of course, noticed that they were taking
extreme measures, and spending millions
to protect the ecology and to safeguard
the animals.

I listened to that Sierra Club editorial
for about three minutes, and I heard
them attempting to tell how the oil com-

panies were destroying the ecology of the
North Slope of Alaska. They made accusation after accusation after accusation.
I listened intently, and then when the
next program came on I remarked to the
people in whose home I was staying that
what had just been presented was rather
odd. I reminded them that I had been in
Alaska for two summers and one winter
and had actually watched what took
place on the North Slope of that country. I told my friends that I could not
find a single accusation in that Sierra
Club editorial that was true—not one.

Naturally they wanted to know more,
and I told them how I had watched the
caribou, animals that did not even know
what a white man was, and had never
seen a work camp before in their lives. I
had actually watched them come through
the work camp, because they had no fear
of us. We could not shoot them, and
we were not allowed to damage the mi-
gration pattern in any way at all. I had
actually watched an entire herd of caribou walk through a Trans-Alaska Oil
Pipeline work camp with no fear of a
human being whatever. As a matter of
fact, I had actually seen them bring forth
their young right on the pad at the work
camp. I had watched those animals
come over and actually settle down right
beside the road, and swimming in pools of

water and ponds and rivers. Man had
never been in this area before, and the
men who were there now were not
damaging the wildlife in any way at that
time, so the caribou had no reason to
fear us.

I have even watched bears walk right
up to a truck that I was driving, obviously having no fear of me, because they
had no natural fear of man in those
areas. Man had never bothered them in
this world of the caribou and the bear.

Thus I was able to substantiate my
argument that there had not been one
single true accusation in the entire three
minutes of that radio editorial. It made
me realize that the American people were
being brainwashed. It became apparent
that the authorities had no intention of
telling the facts about Alaska and the
Pipeline, and this bothered me because I
very much wanted the American people
to know the truth. I wanted them to
know what was really happening at Prudhoe Bay. I wanted them to know that
America needed leadership that would be
honest with its people.

Let me state clearly that I am in sym-
pathy with some of the aims of the ecologists. I am a lover of the outdoors and
I certainly agree that species should be
protected. However, I think that the
matter had reached a point of absurdity

when $2 million dollars was spent rather than removing the nest of a falcon. In view of the many other frustrating experiences which the oil companies endured, it is very difficult to reject the conclusion that there were deliberate efforts to cause costs to be raised to the highest point that was possible. We shall substantiate that view as we proceed.

Chapter 9

How About An Outhouse for $10,000

(Extra for the Mercedes Engine, Of Course!)

There were some rather odd paradoxes in the matter of toilet facilities at Prudhoe Bay, and although the subject matter of this chapter may seem a bit crude (even though we have discussed the subject as delicately as possible), it is necessary to show to what extent excess expense was forced upon the oil companies, adding daily to the tremendous budget overruns.

At first it was official policy to hire only men on the pipeline, it being thought that the rough and tough life that was common to the pipeline was not for

women. Then that policy was changed
and a number of women, of every age,
were allowed in as workers. There were
no separate facilities for women for the
first few months, so they had to live in
the same dorms as the men, even using
the same bathrooms.

The dormitories were built so that 52
men were in a unit, there being two to a
room, and the restrooms were in the center. I admit it was somewhat of a surprise to me one day to be in the bath-
room and notice under the next door a
pair of lady's shoes. Apparently it did
not embarrass the lady, for she seemed to
act as though that was a most natural
thing for her to be there, to come out to
wash her hands, and then to go on her
way. That was life on the Pipeline for
some time. You never even knew if the
person in the shower stall beside you was
a man or a woman.

Obviously sex was an important subject
at the Pipeline, even when women were
not present. There were some places,
such as storehouses, where you simply
could not look at any point on the wall
without sex symbols being depicted. I remember one day when I was out with
Senator Hugh Chance and our truck broke
down. We had to wait a couple of hours
in a room that was about 70 feet long and
40 feet across. Both walls were com-

pletely papered with nudes, from all the pornographic magazines that found their way to Prudhoe Bay. We were there for two hours—there was nowhere else to go, and about the only way to avoid seeing the pornography was to lie down and go to sleep.

Eventually the women had their own dorms, but one could not help sensing that they were not especially embarrassed by sharing the common facilities. The men, in general, had little respect for the women, even though some were decent and respectable. The building of these extra dorms was, of course, an extra expenditure that had not been anticipated at the beginning of the project.

The environmentalists had some weird ideas regarding human waste disposal while the Pipeline was being constructed. The oil companies were forced to use a Hercules aircraft to remove human waste off the slope to Anchorage. The Hercules is a massive four-engined aircraft, able to cart something like 48,000 pounds as a usual load. The tail opens up and the cargo can be loaded. Human excreta was loaded onto Hercules aircraft and tanked all the way to Anchorage, 800 miles away.

As it happened, the sewage system was not operating correctly at Anchorage at that time, so this excreta was dumped into the ocean. The sewage at Anchorage went directly into the inlet because the sewage system was not working effectively—there had been some massive problems with it, and the scheme itself was abandoned for a time.

At first thought, the use of a Hercules for this purpose seems incredible, but it is true. The oil companies were forced to take that human excreta from the slopes where there was virtually nobody living. Out there the excreta could do nothing but fertilize the ground, without having an effect on human beings at all, but the companies were forced to haul it down to Anchorage anyway. Well-placed officials made it clear that it would have been far more sensible to set up designated areas where the waste could be dumped, and then all that would happen would be that the grass would grow, the caribou would be fed, and there would be no problem of the sewage being dumped into the inlet at Anchorage. Obviously large numbers of people could be affected by the foolishness of disposing of the waste in the way it was done, but the ecologists were adamant.

This was not an isolated incident. There were other places where the human excreta had to be tanked into Hercules aircraft and taken away from the slope— another example being in association with the building of the Gilbert Lake Camp and the road in that area. One estimate was that it cost $6,500 for one round trip by Hercules to get rid of a load of human excreta. Anchorage was not the only place that benefitted from this type of unwelcome deposit: Fairbanks was another, and it is now said that Fairbanks has the most unsanitary landfill in all the world. This waste was dumped into the river nearby, and it simply washes off.

There were loudly voiced protests that these were deliberate ways to make the oil companies spend large sums of money unnecessarily, and the fact is that evidence suggests there is much truth in such assertions. The money that was wasted is almost incredible. Millions of dollars were being spent on mobile sewage treatment plants so that the human waste could be carted from the drilling rigs and camps. Samples were sent to the State authorities regularly, and they insisted that tests were run to make sure that the ground itself was not contaminated with human excreta—excreta that, after all, would simply make the grass grow.

The controls were not limited to the Federal government, for State regulations were also very stringent. One of the regulations specifically states that all incinerators shall meet the requirements of Federal and State laws and regulations, and maximum precautions will be taken. Human waste is included in the discarded matter that must be gotten rid of, and it is specifically stated that, after incineration, the material that is not consumed by the incinerators shall be disposed of "in a manner approved in writing by the authorized officer." The State officials decided that the bacterial tanks in use that were fed with air were not acceptable. So they got some long white paper, set the bacterial action going, and whatever was left over was picked up on the paper that was rolled slowly through the water. This then went into a little incinerator and was burned. The ashes were taken to the sanitary landfill and they were buried.

In other words, the incinerator was really a kind of an outhouse. A diesel rig was used, and for a 35-man camp approximately 50 gallons of diesel were used each day. Remember, this was at a time when there was supposed to be a diesel crisis, and it was very difficult to get diesel fuel for jet planes. Because of manipulation, diesel was hard to obtain, and yet the State insisted that human excreta be burned up in this way. A Mercedes Benz engine was used, and it took approximately 350 gallons of diesel each week to run it.

As one highly respected official said,

"Those Mercedes Benz engines are burning up 350 gallons of diesel every week just to get rid of human waste which the tundra desperately needs." He went on, "They do things like this in a very wasteful manner—such as using up 100 pounds of propane every three days, just to get rid of some human turds—why, ever since life began you simply put it on the ground and it makes the grass grow. Now suddenly it's supposed to kill the grass—I haven't figured that one out yet."

These things are not hearsay. We are not giving rumors or secondhand material.

Let me tell you about one day I personally investigated a $10,000 outhouse. I had set out one day to go out to a work-site, riding with one of the engineers at Franklin Bluffs Camp. I often got in the trucks and rode all day with one or another of the men, in order to be out where the men were. I wanted to be right on the work-site and to find out as much as I could. I was anxious to share with men in real life situations and not simply to see them on my terms. I had executive privileges, and so I was free to come and go as I liked.

I enjoyed the drive out with this engineer, and, of course, we talked at length about many aspects of this fantastic project. The engineers are often proud to tell you that they are engaged in what is believed to be the greatest engineering project ever undertaken by man, in all the history of the world. They believed in what they were doing, but over and over again they were frustrated by the limitations set upon them, by the endless regulations that are so often needlessly enforced. They believed there were deliberate efforts to slow down the project and to escalate its cost.

So on this particular day I was riding with this engineer out from Franklin Bluffs. There was one of those outhouses out on the job site, in the middle of nowhere.

I turned to my engineer friend and I said, "Hey, you mean they even have to have privies up here in the middle of nowhere? That tundra surely needs manure—it would be a good idea to fertilize it. After all, there are lots of animals coming through here, and I haven't heard of anyone trying to put diapers on the caribou yet."

"Well," the engineer answered, "We don't dare drop any waste up here, even though the men will be here only a few weeks. According to the government officials we must not fertilize the tundra, because that might not be good for it. We've been instructed to put outhouses every so many miles up and down the haul road of the Trans-Alaska Oil Pipeline, and to have one for every so many men."

I looked at him, hardly able to believe my ears. Here we were out in the middle of nowhere, and intelligent people, products of Western Civilization in the 20th Century, were seriously suggesting that high quality outhouses must be put up at regular points. I chuckled and said to the engineer, "Hey, that's interesting—how in the world could they have an outhouse out in the middle of nowhere? After all, everyone that goes in it would freeze."

"No," the engineer answered. "Reverend, you won't believe how much that outhouse costs—the very one you're looking at over there."

I looked across in the general direction he was nodding to. "Well," I said, "we used to build outhouses for nothing—we'd use scrap lumber on the farm." The engineer nodded. "Yes, that's what you'd do back on the farm, and that was the sensible thing to do, but we're not allowed to do that up here. We can't even dig any holes in this tundra to put an outhouse on—we are told that that would destroy the ecology. The regulation is that we must have these *special* outhouses hauled in."

I was finding it hard to believe my ears. Here was a highly intelligent man telling me that officialdom was of such a nature that apparently huge sums of money must be spent on these "special" outhouses. I turned to the engineer and asked, "Well, what's so *special* about them?"

He answered, "The first thing that is special about them is that they cost **$10,000 each.**"

I looked at him in surprise. "Wait a minute, sir," I interrupted, "You're talking about an outhouse—you're not talking about buying a Mercedes Benz."

Then he gave me a smile. "As a matter of fact, that outhouse has a Mercedes Benz diesel engine on it. When I said $10,000, I didn't mean the engine—that's extra, of course."

"Come on now, explain it to me. What's all this nonsense you're trying to put over?"

The engineer assured me it was not nonsense. He said, "You see, that's an entire self-contained incinerator unit, and if ever you saw the black smoke coming out of the stack of that thing, and then you smelled the aroma, you'd really know what contamination was. It surely is contaminating the air, and the whole ecology, too."

"How does the incineration process work?" I asked. "Well," the engineer answered, "When a man does his business in that outhouse, it goes down to the bottom, and that diesel engine automatically cranks up. By electrical and other means it completely incinerates everything." He pointed to a pipe that came out from the outhouse. "It shoots out that pipe up there, and as a result it's not supposed to contaminate anything. Well, I can only say it certainly contaminates my nostrils all the time."

Right then I knew that my own nostrils were being contaminated in no uncertain way, and while I was there I always knew when someone was "Doing his business." I found myself annoyed at the idea of a diesel engine automatically cranking up for such a purpose. I must confess, too, that whenever I go to the gas pumps and buy fuel, I remember that my own pocketbook has been contaminated—contaminated by those outhouses at $10,000 each, *plus the cost of the Mercedes Benz engine, of course!*

$10,000 (plus) for an outhouse with a Mercedes Benz engine thrown in? Just because they didn't want to fertilize the tundra! This was bureaucracy gone mad. For what purpose? We shall answer that question as we proceed.

Chapter 10

One Law for the Rich, Another for the Poor

We've talked about the two-million dollar falcon's nest, and the $10,000 outhouses. There were many other similar incidents—they can be multiplied, and taken together, they involved a huge sum of money.

Another method to add to the price of the pipeline, and again to the price that you the individual will pay at the gas pump, was the almost incredible use of fines. On one occasion a vehicle with sightseers on board ran off the road to let a truck go by. No damage was done —there was nothing off the road, just the tundra. Remember that it would take an ax to break through that tundra. Nevertheless, there was a fine of $10,000 levied because that vehicle ran off the road. Of course, it was not the sightseers that got fined, but the ARCO company.

People living in the lower 48 will find it hard to believe that such practices continued, but they surely did. Another case was where a pickup truck drove into the river to turn around. A security guard had locked the gate, and so this was the way that the driver solved his own problem. Again the ARCO company got fined $10,000 for not making an adequate turn around. They hurt nothing driving their vehicle into the river, and it is really impossible to figure out why they should have been fined—but fined they were.

The amounts of these fines were announced in the paper very often, and there would be a small write-up. It didn't make big news, for the policy seemed to be to keep these matters in low key. It is ultimately the poor guy who buys gas for his automobile that pays those fines of $10,000 and more—for the most trivial offenses against the huge number of regulations to which the oil companies were subjected.

Not only were there very heavy fines, but also they dragged the work out. One section of road was supposed to be a five-week project, but because of government meddling, it was about 3 months before it was finished. The government tinkered with the administration, fined the company, and stopped them in all sorts of ways. They told them what they could and could not do, when they could work and when they could not. At one time there were 22 government monitors working on that one section of road. They came from such departments as the Department of the Interior, the Department of Fisheries and Game, and the U.S. Geographic Coastal Survey. Most of them were Federal workers, but some were State workers also. Those 22 workers were running around surveying the same stretch of road at the same time, day after day. While that stretch of road was being built, some 18 fines were levied—in a three-month period. Every one of those fines was for at least $10,000.

The company that had the contract for that stretch of road ran over their estimated budget by about $5,000,000. The cost overrun almost broke them, and the ARCO company had to come back and reimburse them to keep them from going bankrupt.

There was no doubt that by the strict enforcement of often ridiculous and excessive regulations, the attempt was being made to bankrupt all the oil companies. Often regulations were changed; a good example of that was when the rules for going on the tundra were altered. It used to be that you could not go on the tundra unless there had been 30 days of consecutive freeze and a specified amount of snow. Then the authorities would issue a permit, and you could go anywhere you liked on the tundra—after all, you cannot hurt it. Then the regulations were changed to make it so that you could not go on the tundra for any reason without a permit. Anytime you wanted to go on the tundra you had to have a specific permit registered with the State—and it would take weeks to get one. Of course, people had to be paid to process those permits.

This new regulation was considered by many people to be absurd, for there were all too many occasions that it was necessary to go on the tundra in the normal course of events—to check out a marker, or to repair a light pole, or for many other legitimate reasons.

The tundra is not easily scored or damaged. You could drive all over it right through the winter and never see where you had driven. You need an ax to break it up, yet the authorities made it essential to get these permits. They were State people because the land is State-owned, not owned by the Federal government.

The same controls extended even to the dumps associated with the camps. One oil company executive told me that there were three State ecologists monitoring the dump where he worked. They lived at Attwood, and there were three of them employed, with no other work than the monitoring of that dump. At that place there is the only certified landfill in the North Slope!

One day these three monitors came to the dump, and someone had dumped some spoiled weiner packs—hotdogs—and of course hotdogs are supposed to be buried. On this occasion for some reason the garbage man had mixed up one of his bags and got the whole bag of spoiled hotdogs and dumped them on the dump.

These three people found the hotdogs, and they fined the company $10,000 for throwing hotdogs away. Their argument was that food should not be thrown on the dump because it would attract bears. The fact was that this was a legitimate mistake, for the company operated its incinerator and a man was paid to burn all that stuff. He just did not get it done that particular day, and so the company was fined $10,000.

The same company executive, who indignantly told me about the hotdogs, also pointed out that it was not permitted to salvage anything from the dumps. Often it would cost large sums of money to freight iron, copper, and brass to the site, but it was then buried at the dump.

Nothing could be moved out, even if it was urgently required, e.g., for repair purposes.

When the fines were levied, there was little the offending parties could do about it. The fines were levied, and the amounts were learned 2 or 3 months after the incident.

There is an old saying, "One rule for the rich and another for the poor." It certainly was true that there was one way to apply these regulations to the employees of the oil companies and another way when it came to the State employees. We've just said that the company was fined for allowing a bag of hotdogs to accidentally be thrown on the dump because it might attract the bears. Yet some of their own employees did worse things with food lying around, and it did, in fact, attract bears. Then *those employees shot the bears, and nothing was done!* No action was taken against them . . . *not even a fine!*

The oil company people were not allowed to participate in hunting or fishing: they were fired if they got caught. A different set of rules applied to the State employees.

Here is another example—ARCO transferred to the State of Alaska the Dead Horse airstrip and camp. The camp itself was sold, but the airstrip was not, it being a gift. The company had put millions of dollars into that airstrip, and it was in fact the finest airstrip in the State. Those who know the facts would agree with that assessment, and would also agree that the airstrip has not been maintained properly since then.

The State authorities sent a tower man to live up there, and he was allowed to keep his wife there. The radio man maintained the radio and there was a mechanic to maintain the equipment. Maybe there were others also—they certainly had a Fisheries and Game man there.

A team of people came to that airstrip, and they would just throw the garbage out their back doors, which was something the oil company employees were not allowed to do. They had to incinerate *all* their rubbish at all times. So it was that the bears got to eating on the back porch where these State officials would throw their garbage, and then the officials themselves killed the bears and flew their hides out.

That was in Prudhoe Bay, and it is widely known that they did what I am saying. The company's environmentalist wrote to the State authorities about it, but to no avail. Those people killed every bear in Prudhoe Bay: there's not a bear to be seen in the oil fields there now. These "outsiders" brought their guns in, shot them, tagged them, and hauled them out. By "tagging" we mean that they were supposedly legally shot, a hunting fee having been paid. Even that was something that was not legal for the oil company employees to do. Those bears were actually pets of the oil field, and they were ruthlessly shot by these employees of the State. There were about 7 bears that lived more or less as pets around the oil fields —7 Plains Grizzly bears, these being a rare breed Grizzly bear. They are a little smaller than the Kodiak Grizzly, with bigger heads and wider. They grow to about 9 or 10 feet, instead of 11 feet which is common with the Kodiak bears.

Bears were commonly seen around the camp. They would go back into the mountains and hide there in the winter months, but they would come down every summer and live in the fields around Prudhoe Bay—until the State people killed them. There was one mother bear with her three cubs living around one of the camps. Nobody had any problem with her—she was regarded as a pet. Another mother and her cub did cause some trouble, and they were put in a helicopter and carried about 150 miles away and unloaded, but they were back in their original camp area two days later. Would you believe it, the company actually got fined for taking that bear and her cub in the helicopter and removing them! Yet State employees killed bears and no action was taken against them.

Things were very different with these State people. They actually killed the cubs, as well as the adult bears, and this was common knowledge. Though the oil company environmentalists reported it, even getting one of the security guards as a witness, no action was ever taken on this entire matter.

The State people concerned did not have to stay long in the area. The tower man could only stay there one year, but then he could go somewhere else, such as Anchorage, Fairbanks, or even to the State of Hawaii.

As we stated above, there is a saying, "There is one law for the rich and another for the poor." At Prudhoe it was quite obvious that there was one law for the oil companies and another for the State.

Chapter 11

The Barges Froze— and Cracked— and Popped

Time went by. Now I had been Chaplain on the Trans-Alaska Oil Pipeline for two years. I had spent two years watching and examining, in constant contact with the men who were planning and then undertaking the construction of this great project.

Now it was all beginning to add up, and here is the way it looked.

In 1971, the oil companies had first proposed the Trans-Alaska Oil Pipeline. At that time the projected cost was $600,000,000. That was the anticipated figure in 1971, but before it could actually begin in Alaska, the government stepped in and said, "No, until more surveys are undertaken, and more guidelines have been

laid down in such areas as the protection
of the ecology, you will not build the
pipeline.''

The nine major oil companies of America had hauled that big pipe from Japan
to Alaska. It is interesting to notice that
the pipe itself had been built in Japan,
because prices were already beginning to
go so high, even back in 1971. By that
time it was cheaper to buy it abroad and
ship it across the water to Alaska. So it
was that an American bank financed a
Japanese steel company for the purpose of
building the big pipe for the Trans-Alaska
Oil Pipeline. While the pipe was actually
bought and made in Japan and then
shipped to America, it had to be stored
from 1971 until 1974 in Pipeyards—in
Fairbanks, Valdez, and Prudhoe Bay—
three sites in Alaska. Then in 1974, the
pipeline began to take shape: the government had issued their permits, surveys had
been made, the ecology had been studied
from 1971 to 1974, and an entirely new
method of building the Trans-Alaska Oil
Pipeline had been devised. At that time
inflation was beginning to cut even deeper
into the American economy. There was
an increasing spiral of inflation in the
early 70's—up to that time the prices remained more or less the same year after
year. When the pipeline began to be ini-
tiated in 1974, the cost estimate was no

longer $600,000,000 (600 million dollars), but $2,000,000,000 (2 billion dollars)!

Moving on to 1976, it was interesting to stand and look back, and also to look forward. In 1971, the figure was $600, 000,000—we needed the oil at that time, but there was no energy crisis. Nevertheless, the country needed oil and private enterprise could produce it. However, the oil was on government-owned land, and so the project was stopped until government had their say. In 1974 the project cost was $2,000,000,000 for the cost of that pipeline. Now we reach 1976, and the oil company officials were saying that, because of cost overruns, the total cost of the oil pipeline would probably exceed *$12,000,000,000.* At that point it was all beginning to add up. I was beginning to realize that there was indeed something in the wind.

There was an underlying force that was attempting to control both the oil companies and the flow of oil. From 1976 on, frustration began to be intensified. Permits were withdrawn, even though they had been issued for the entire time of the construction of the Trans-Alaska Oil Pipeline, and had been promised as such by the Federal government. Now I was watching as one after another they were withdrawn in an attempt to frustrate the entire project. Regulations were being

intensified—there had been plenty of time
in two years to update the regulations by
which the government controlled the whole
operation, in such matters as the protection of the environment.

I remembered that first book dealing
with regulations that I had taken to my
dorm room in 1974. Even at that time I
had read through it very carefully and
wondered at what I read—private enterprise was building this immense pipeline,
and yet was being told what to do in
minute detail, having to get specific permission at all sorts of points from the Federal government, even though that government was not putting one penny into the
entire project. I was watching as their
permits were being withdrawn and even
more stringent regulations imposed.

It indeed seemed that the Federal government did not want the oil to flow. The
oil was found on Federal and state lands
north of the Brooks Mountains, and most
of the land was owned by the Federal
government. 92% of all the land in Alaska is owned by the Federal and State governments. Only 8% is owned by individuals, so the oil is on government-owned
land. So it was that the oil companies
were told what they could do, in very
great detail.

I had always thought of the government
as having been elected by the people, for

the people, and of the people, so surely the government would want what was best for the American people. Surely we have not lost sight of the fact that private enterprise has made this nation so great and prosperous. That has been so since the time that our forefathers devised the method of incentive to allow private companies to develop and produce. This land in Alaska was owned by the Federal government—therefore, is not this the land of the people of America? Did they really *not want* the fuel to be produced? If that was the situation, why? There was *supposed* to be an energy crisis.

Then I remembered that Mr. X had said that the oil companies had been allowed to to produce oil for the Trans-Alaska Oil Pipeline from only one 100-square-mile area of this North Slope of Alaska, and I remembered that the North Slope of Alaska includes many times 100 square miles. Mr. X had said that all of the land north of the Brooks Mountains included many pools of oil—it was there in vast quantities beneath that North Slope. Nevertheless, private enterprise and the oil companies of America are allowed to produce from only one of those pools. They have been deliberately limited to one 100-square-mile area.

Then I remembered that *"precious"* tundra—that seemed to be all I could hear

about on the news and from the ecologists
. . . the cry constantly was, "Preserve the
tundra!" . . . the tundra was so *precious*.
Yet I actually watched them lay large
areas of styrofoam for insulation under
the road, a road that was nothing but
gravel. I watched them bring in truckload
after truckload after truckload of large
sheets of styrofoam, and then they would
lay them straight onto the tundra, then
the gravel would be put on top of that.
They would lay a gravel pad on top of
that styrofoam just to keep the ground
from thawing and to preserve the tundra.
I watched reseeding taking place after they
had laid the pipe. I thought of those
men who were literally fired because they
happened to drive a bulldozer out on the
tundra, off the road that had been built
—a road that was actually a road laid out
across the bare North Slope.

I remembered that I had watched the
caribou who had never seen humans before, and that I had watched the bears,
bears that did not know that they were
supposed to be afraid of us, walk right
into the camp. I knew that their migra-
tion paths had never been disturbed. Even
the wolves had no fear of man in these
areas.

It seems rather strange that today,
about three years after the oil pipeline has
been completed, that its construction did

not destroy the environment or disturb the
tundra, or other aspects of the environment in any major way whatever. Let us
summarize a few facts that we have already presented, and some others that are
just as relevant. I thought back to those
two falcons, falcons that could not be disturbed while they were nesting, and so
$2,000,000 had to be spent to reroute the
road rather than disturb them at that time
—$1,000,000 per falcon.

Next, I could never forget that large
flotilla of barges that were brought each
year from the West Coast of the lower 48
states, bringing all the supplies and equipment necessary for the Prudhoe Bay oil
field. Entire buildings and other constructions had been assembled in the lower 48
and placed on huge barges and floated by
way of the Pacific Ocean through the
Baring Sea, then into the Arctic Ocean,
and eventually across to Prudhoe Bay.
Each year one of the highlights was when
the flotilla of barges came in. They
brought everything, from the big pump
stations to the flow stations to the pipe
itself. They brought in vehicles, dormitories, and everything necessary in the way
of large construction equipment, such as
drilling rigs . . . and on and on. They
brought in everything that was needed for
the work of producing oil from the fields
at Prudhoe Bay.

Then in 1975, the weather just simply
did not cooperate. That flotilla would
have to wait until the Arctic ice had left
the ocean. The flotilla would usually
stand for weeks at a place called Wainwright. They would wait for the ice to
move at Barrow, and then they would
have only a few days in which to get out.
We would hear the message, "The ice is
moving! The wind is moving from north
to south—there's a shifting!" So they
would move out into Prudhoe Bay.

In 1974, the fleet had plenty of time to
get around Point Barrow and into Prudhoe Bay, and to get back again to the
lower 48 in protected iceless waters for the
winter time. However, this year (1975)
the weather simply was not cooperating,
and every single hour was precious. Every
moment had to be counted. Finally the
ice broke just long enough for the flotilla
to come around by Wainwright and Point
Barrow. Then it arrived at Prudhoe Bay,
but something was wrong—the ice was
barely staying off shore, so the flotilla
did not have time to get back out. The
ice closed in again, the wind was not fav-
orable, and soon it was clear that the flotilla of barges and the tugboats that
brought them in would be stuck at Prudhoe Bay for the winter—they could not
get out again.

This presented a problem. Before the

Arctic Ocean and the Beaufort Sea closed in again, somehow these barges had to be lifted out of the water and brought onto the land. However, the water is very shallow close to the land at Prudhoe Bay, and the barges were a long way from shore.

The equipment was brought in piece by piece, but then there were the expensive barges owned by the companies, and the tugboats that brought them in—how could *they* be saved? There was really only one way, and that was to build a dock. Why not? Put gravel out into the ocean and dock them on dry land, so that the ice would not crush them in the winter. Then I watched, knowing that time was precious. The Federal bureaucracy cares nothing for time, and seems to care nothing for private enterprise. The fact that they had millions of dollars in equipment tied up there, sitting out in that water, mattered little.

The water was gradually freezing in around the barges, and it would crush all that equipment. While they were deciding what to do, *that is exactly what happened!* The ice closed in around those big barges. They were able to save the tugboat, but the barges were left in the water, because there were some microorganisms on the bottom of the Arctic Ocean at that point, and the ecologist insisted

they must not be destroyed by the building
of a gravel dock out into the water to
the point of the barges.

I watched as the big equipment was
brought in. They were actually outfitting
bulldozers so they could ride the bottom
of the ocean and literally go up to the
place where the barges were and pull them
in. I saw huge nylon lines, bigger than I
had ever seen before, brought in. They
said that the big 'dozers would literally
pull the barges in, but then—*NO!* Surveys would have to be made . . . it would
have to be found whether they were going
to destroy any of those microorganisms,
and the little, minute fish that swam on
the bottom of the Arctic Ocean for only
a short period each summer. The argument was that by taking those 'dozers out
into the middle of the ocean, for only a
few hours, especially equipped as they
were to pull those big barges in, they
might somehow—*just might* destroy those
microorganisms! In this area, even though
it runs for hundreds of miles, we did not
dare disturb the ecology, and surveyors
must make their tests before a dock could
be built or 'dozers could be used to bring
the barges in.

I watched as they stalled, and stalled,
and stalled for time . . . until they had finally stalled long enough! The barges
froze, and cracked, and popped. The big

steel plates were literally destroyed, and millions of dollars worth of equipment was crushed by ice—*Why?* Could it be that the government did not want that flow of oil? *Could it really be that there is no energy crisis, except the one they want to produce?*

Then came that $10,000 outhouse to which we have already referred. Why $10,000 for an outhouse (just to prevent the tundra from receiving needed fertilizer, . . . and remember, *nobody tried to put diapers on the bears and caribou)*? No other company in American had to pay that price for an outhouse! They do not require such extravagance in our polluted population centers—yet there was no pollution, except for the pollution that was coming out of the smokestacks of those same $10,000 outhouses. You could smell it for miles if the wind happened to be blowing in your direction.

Further, I noticed sewage systems were having their permits withdrawn—from Galbraith Lake all the way to Prudhoe Bay. All withdrawn, even though they had been issued for the promised life of the Trans-Alaska Oil Pipeline construction. Now within 9 to 12 months before the completion of the project, sewage systems were being removed—*Why?* The water coming from them was perfectly pure. They had met all regulations and standards. They had been approved and permits had been issued. Yet orders were now being issued for these sewage systems to be removed, and new ones, at exorbitant costs, were being brought in for one more year of the life of the construction of the Pipeline. All this was because of Federal and State government orders.

Was it an attempt to frustrate the construction of the Trans-Alaska Oil Pipeline?

Then came 1976 and the last six months of the construction of the line. Here I was as Chaplain, in the midst of what appeared to be a planned frustration. If I may use the word without being misunderstood, there was apparently a plot to keep that oil from flowing. At the same time, all across America, there were lines of people standing and waiting for fuel. There was talk of rationing, and yet there was plenty of oil in Alaska, and apparently there was a frustration to prevent it from being used. The oil companies were doing their utmost. With all their power they were attempting to complete this pipeline and to supply oil for the people of our nation. Private enterprise has always done that from the beginning of this great nation.

Now that we had come to the last six months of the Trans-Alaska Oil Pipeline construction, it seemed that everything went wrong. Suddenly there was another turn. Someone had said that the welding on the big pipe was faulty, but how could that be? I had watched day after day. Almost daily, in order to rub shoulders with the men as much as possible, I had driven up and down that long stretch of pipe where they were welding it together. I knew the men, the welders, and the other men who were laboring there. Many of them were in my worship services week after week, in the seven camps from Galbraith Lake to Prudhoe Bay. These included the men who were actually doing the welding, as well as the men who were X-raying the welds.

I asked them, "Are those welds on that pipe faulty?" And then, only months from the completion of the whole project, there was this possibility that the whole big pipe would have to be redone, from beginning to end. Where it went under the river bed it would have to be dug up. Can you imagine the destruction of the ecology if such a thing was to take place? The suggestion was that it be dug up where it had already been laid in the ground under the streams. At this time of the year that would have been almost literally an impossibility, for it would have destroyed the fish streams and the breeding grounds—that was what the ecologists hollered. They insisted, "It can't be done now—you must wait. After all, we don't dare touch those streams at this time of year."

It was clear that a deliberate attempt was being made to stop the flow of that oil, to prevent the whole project from being brought to a successful conclusion—it seemed that the intention was that it would never be produced. The plan became increasingly clear and the tension increased every day.

The company that was X-raying the big pipe was accused of duplicating film. The charges were simply not substantiated. To the best of my knowledge, *there was not one single leak* after the oil went through, but you never heard it told later that all those millions spent at that time *were spent unnecessarily. That received no publicity!*

It became clearer that all of this was somehow planned. For two to three months all we heard was, "Faulty welding!" The word went out all across America that the pipeline had to be stopped—and even dug up. America was told that the oil would leak out onto the ground and would destroy the "precious" tundra. The news media proclaimed that this would be the biggest oil spill ever known on the face of the earth, and it must be stopped. Three years later you have heard of no oil spills, except those which were produced by people who attempted to sabotage the pipe after the oil actually began to flow.

You find no streams north of the Brooks Mountains with oil flowing into them because of the oil seeping from the ground where the pipe was laid. No, because there were no faulty welds in those pipes.

I am not merely giving an opinion—I had it from a thoroughly acceptable witness, as we shall see in our next chapter.

Chapter 12

Those Welds Are Not Faulty!

Sometimes it seems that things happen just by chance, but I do not accept that. I am a great believer in Divine Providence —that God can and does guide those who seek to follow Him. Thus I believe that what I am about to relate was part of that Divine overruling.

Let me start at the beginning. I have said that as Chaplain on the Pipeline I was responsible for seven camps—all of the camps north of the Brooks Range from Galbraith Lake Camp to Prudhoe Bay. Each day I had a worship service in a different camp, seven days a week. The camps were approximately 35 to 40 miles from each other.

This meant that I traveled each day from camp to camp, and upon arriving I would check in with the registration desk. The lady at the desk would attempt to

give me a room by myself, if possible, because of my position as Chaplain. I had
to do a great deal of counseling, and
clearly it was desirable to have a room
where privacy was possible. On this particular day, I arrived at Franklins Bluff
Camp, and the lady at the desk said,
"Chaplain, I would very much like to give
you a room by yourself, but we are just
full up today and it is not possible."

I said to the lady, "Thank you, I understand the situation. I do not mind at
all sharing a room with someone else."

She handed me a room number on a
slip of paper, as she normally did, and I
thought very little about it as I walked
down the corridor toward the dorm section. I walked into the 52-man dorm,
down the hallway, and started to enter the
room to which I had been assigned.
There was an immediate protest from
someone inside the room I was entering.
A gentleman came to the door and said,
"I'm sorry, but you can't stay here."

I replied, "Sir, I'm very sorry. I didn't
mean to intrude."

At this point the man walked out of the
room, and said "You'll have to go back
to the desk and get reassigned."

I turned and started to walk away, and
as I did he said, "Hey, by the way, who
are you?"

I answered, "I'm the Chaplain with

Alyeska Pipeline Service Company, assigned to this camp."

The gentleman smiled and said, "Well, Chaplain, I think you just might be interested in this, since you are a Reverend."

Then I asked him, "Well, why did you not want me in the room? After all, I don't want to intrude any place where I should not be."

The gentleman explained that he was appointed by Alyeska and the Federal government to examine the so-called "faulty welds." The claims that the welds in the big pipe were faulty had been spread all over the country by the news media—in the newspapers, on radio, and on television. It was put out for the whole world to know that the welds were bad. Every so many feet the pipe must be welded, and the Federal government had claimed that many of the welds were faulty, and that as a result there would be leaks when the oil was flowing.

Remember that the pipeline was both above and below ground. From Prudhoe Bay to Valdez was approximately 800 miles. The estimate was that approximately half of the pipeline in that distance was under the ground, and the other half was above the ground. To check the faulty welds, as the Federal government wanted, would mean the digging up of virtually hundreds of miles of the oil pipeline. Each of these welds was supposed to have been X-rayed prior to the pipe being laid in the ground.

Indeed, the problem was even more serious than simply going underground. Much of the pipe was actually underneath river beds. This, therefore, would have meant literally multiplied millions of dollars for the pipeline to have been dug up and X-rayed again. All the X-rays of the welds prior to the pipe being laid in the ground were undertaken by a firm that was subcontracted by Alyeska.

Another highly relevant fact was that these instructions to investigate the welds came up only some six to nine months before the projected date of oil flow. Obviously such an undertaking would cause great delay, and the costs would be enormous. The claim was that the company that was supposed to have X-rayed these welds had duplicated their film, and in so doing had cut back on their own costs, but had not done the job properly. It should be pointed out that when every joint of pipe was put together, the weld had to be X-rayed, and the company had films to prove that the X-rays had been carried out.

With that background, let me go back to the gentleman in the room who had protested my entrance. I still did not know the man's name, but now he smiled and said, "Reverend, come on in."

As we walked across the dimly lit room, I noticed a light table on which were placed many strips of film. He explained that these were the films which represented the welds on each joint of the big pipe. This four-foot pipe that was to carry the crude oil from Prudhoe Bay to Valdez is the largest diameter pipe ever constructed for the carrying of crude oil.

As the gentleman pointed to the light table, I remarked, "Sir, I know nothing about X-raying the welds on a big pipe. Would you please tell me what all this is . . . and why it's so secret?"

The man said, "Chaplain, haven't you heard about the faulty welds on the big pipe?"

I said, "Yes, sir, anybody who listens to the radio or watches TV or reads the newspaper has heard about that."

He said, "Chaplain, my purpose in being here is to examine those welds." He continued, "Sir, all of this is classified."

I asked, "Do you mean that no one is supposed to see these films?"

He responded, "Chaplain, until this matter is settled, it could be very drastic— it's of national importance."

The gentleman was cordial by now, and he took considerable pains to explain what he was involved in. I asked him what on the films would show whether a weld was good or bad. He took a picture and pointed out a good weld, then put beside it a picture of a bad weld. The bad weld appeared to have bubbles internally.

I asked, "Do you mean to tell me that an X-ray can pick up a bubble inside a piece of metal?"

He replied, "Yes, because the type of X-ray that we undertake is done with radioactive material."

As I compared the pictures, I could see a dark crusty area, and to a normal layman, it appeared to have what looked like a bubble. On the picture, a good weld looked exactly like a good weld on the outside of a piece of metal would look. Remember, I'm talking about this as to how the thing would appear to a normal layman.

The gentleman told me that what had to be proven was whether or not each weld had actually been X-rayed. I realized as he talked to me that this was very important, and I spent a great deal of time going through the details with him as he explained various points to me.

It should be stressed that I was shown these X-rays without any coercion on my part. I did not so much as ask to see them, and he at no time asked me to keep secret what he showed me, or anything that he told me. He did allow me to share the room with him that night, and we talked at considerable length.

In the course of our lengthy conversation it became very clear that this gentleman believed that the whole investigation was unwarranted, that there was no truth to the claim that there were faulty welds, and that it was costing the oil companies millions of dollars for this investigation. He had already been through most of the films, and he had simply not come up with evidence to demonstrate the validity of the claims that had been made relating to the welds.

One other point of background is that this gentleman told me there had been an agreement between Alyeska Pipeline Service Company and the Federal government to appoint him as examiner for these welds. It had been mutually agreed that his decisions would be accepted by both sides. Some Federal inspector, hidden in anonymity, had claimed that the welds were faulty, but here was the expert, mutually agreed to by both parties, insisting that the Federal inspector's claims were false.

Sometimes in these matters of high policy, sacrifices are made, and there are even those who become scapegoats. The company that had been challenged as to its integrity in this matter of the welds was actually dismissed by Alyeska Pipeline Service Company, and paid off. Another company was appointed to continue the work, and thus a compromise was reached. This was reported as being a face-saving operation, but in fact the gentleman with whom I shared the room that night made it quite clear to me that basically the charges had no substance.

During the evening we talked at length about what was happening as this mammoth project was nearing its completion. It again appeared that somewhere underlying the total picture was an attempt by the government to postpone the flow of oil. I was left with the clear impression that government intervention was quite deliberate, in an attempt to lead the oil companies to financial chaos, even to their bankruptcy, and ultimately to the nationalization of the oil industry. We shall elaborate as we proceed in a later chapter.

After I left that man the next day, I kept thinking about the things I had seen and heard, and I attempted to put the pieces of the puzzle together. Time went by, and it was later proven by the actual flow of oil that *the welds were not faulty.* To my knowledge, there was no leak that developed in the pipe at any point as a result of a faulty weld. Nevertheless, the fact is that Alyeska Pipeline Service Company was instructed by the Federal government to dig up certain points of the pipe at extreme expense, to re-X-ray the welds, and to re-lay the pipe. Alyeska had no option but to obey, so they did it.

At this point I was told by Mr. X that cost overruns were going to bring the total cost of the pipeline to $12 billion dollars. Remember that originally the pipeline was supposed to cost $600 million . . . then $2 billion . . . but $10 billion dollars *extra* cost as an overrun?—*five times the original estimate! Why?* What is the underlying motivation? What absurd policy is the Federal government pursuing? Why are you paying approximately $1.50 a gallon at the gas pump now?

One reason was that there was a concerted effort to ruin the oil companies, bring them to bankruptcy, discredit them in the eyes of the people, and ultimately nationalize the oil industry.

That became even clearer as those rich oil men from Saudi Arabia, as well as the bankers from the lower 48 states, began coming to Prudhoe Bay in large numbers.

Chapter 13

Why Are These Arabs Here?

Now I was deeply suspicious. I found myself going over the conversations I had with that gentleman, time and time again. In my mind's eye I saw bubbles on X-ray photographs, and I compared good and bad welds. I went over and over the things he had told me. I became convinced that, to quote an old saying, all was not right in the State of Denmark.

Then I remembered something else. In my mind I went back to the conversations between Mr. X and Senator Chance, conversations in which I had participated. That had been one and a half years prior to this time, but suddenly I saw tremendous developments relating to some of the things Mr. X had said at the time. I decided I would put some answers to them.

What follows is an approximate recall of the questions and answers between

Senator Chance and Mr. X, one and a half years earlier. If you like, this is the good old "flashback" method. The questions and answers went like this.

Senator Hugh Chance had asked, "Mr. X, how much oil is there on the North Slope of Alaska?"

"Senator Chance, I'm persuaded there is as much oil as there is in all of Saudi Arabia."

"Then, Mr. X, if there is that much oil there, there is not an energy crisis."

(Mr. X's only answer was a smile, implying that Senator Chance had hit the nail on the head.)

"Mr. X, what do you think the Federal government is really out to do?"

"Senator, I personally feel that the American government wants to nationalize the oil companies of America."

"Then, Mr. X, if you are so convinced of that fact, have you calculated how long you can remain solvent with present Federal control?"

Mr. X was reluctant to answer at first, but then he looked at Senator Chance and said, "Yes, we are so convinced that in fact we, as oil company executives, have made that calculation."

"Then how much longer do you think you can remain solvent?"

"Until the year 1982."

"Then, if what you say is true, why

don't you oil companies warn the American people of what is going on? After all, it is your neck that is at stake."

"Senator, we can't afford to tell the truth."

"Why not?"

"Because, Senator, the Federal government already has so many laws passed, and regulations imposed on us as oil companies, that if they decided to enforce these rules they could put us into bankruptcy within six months. Sir, we don't dare tell the truth."

In passing, we point out that in our later chapters we shall explain how all this ties in with the *apparent* millions of dollars in profits made by the various oil companies today. There is an explanation, and it is mind-boggling!

That was the conversation, virtually word for word, as I remember it. The conversation cannot be denied. Senator Chance and I were both there, and we publicly and privately made it clear that the conversation did take place, just as I have recorded it.

Now I was in an unexpected situation. Here we were approaching the end of the time on the pipeline, and there was a story that must be told. Mr. X had understood one and a half years previously that the American government was out to nationalize the oil companies. He had seen it long before I did, but now I understood that too. Should I remain silent? (And even if I talked, would anybody believe me?) Should I be prepared to open my mouth, because I, as a true American, believe in the free enterprise system? Would there be danger, maybe even physical danger, and would there be attacks against my spiritual ministry if I did open my mouth as to the facts that were taking place all around me?

I have always been one prepared to accept a challenge. I knew that I had no choice. I had no option but to do what had to be done—to do my part to inform the American people of the dramatic attempts that were being made to bring the oil companies to their knees, to a state of bankruptcy, as one of the necessary steps towards the socialization of the great Republic of which I am a proud member.

From that point on I began to pry into everything I could, to find out all the facts that were relevant. I was interested with a *new* interest that I had not previously had. I was a man with a mission.

It might be worthy to note that I was the only Chaplain on this Northern Sector of the Pipeline, and therefore I was the only one who would have access to this particular information. Other Chaplains on the Pipeline would not even have known what I had access to. Therefore they would have no wish to report, either through the media or by such a book as this. I want to make it very clear that in no way am I challenging the integrity of others who were Chaplains at other areas of the Pipeline.

As I mentioned previously, I noticed that permits which had been issued for the life of the construction of the Pipeline were now being withdrawn.

One day I walked into the office of one of the engineers, and he began to show me what was happening at Happy Valley. Before long I found out that this same story was being multiplied up and down the Northern Sector of the Pipeline. There was a lengthy manual published which listed all the permits. I had reviewed it at the beginning of the construction phase of the Pipeline, and I remember very clearly that the words were that these were to be the rules that were to be followed by everybody for the entire construction phase of the Pipeline.

Now we were within nine months of the completion of the Pipeline and of oil flow. This was the status as I was in the engineer's office that day. As soon as I walked into the office he began to say, "Chaplain what do you think of this sort of nonsense? Here the Federal government is instructing us to change the entire system of sewage that we have in this

camp. We are a few months from the
end of our time here, and the system
we've got has proved perfectly satisfactory. If we do what they tell us to do,
it's going to run into a fantastic cost, and
the whole thing will be left here when we
move out in just a few month's time.
Have you ever heard of such nonsense?
What do you think is their purpose? Why
would they want us to remove one system
that they approved only a relatively short
time ago? Now they've decided that that
system is not satisfactory and we must
have this new one."

I was flabbergasted! "Are you telling
me that the system that has only been in
for nearly two years, is now so faulty that
it must be replaced and won't do for the
few extra months we are to be here?"

"Yes, that's exactly what I'm saying. I
find it hard to believe—there's something
wrong somewhere. Sometimes these government regulations are just about impossible to understand. But for us to tear
down and haul out our present system
would involve a fantastic sum of money.
Then we've got to actually rebuild this
new sewage system, bring it in, put it up,
and there is absolutely no point in doing
it. The system we have is perfectly satisfactory. It almost seems as though the
government is doing its utmost to slow
down the development of the Pipeline, and

maybe even to make the costs as high as they can. What do you think Chaplain? Are they trying to break the oil companies, or delay the flow of oil? What do you make of it?"

I looked at him, and then I asked, "What do you think yourself? Do you think the new system is justified . . . is there something wrong with the old system?"

"No, Chaplain! There's nothing wrong with the old system. The water that comes out from that system after it's been treated is so pure that you could drink it. There's absolutely no reason at all why the old system should be taken out. Nor does the water hurt the ecology—it's just good, ordinary pure water. This whole business is utterly ridiculous, and what's more, there are a lot more withdrawals of permits taking place up and down the Pipeline. I wish I knew what was going on."

"Yes, I wish I knew what was going on, too," I answered quietly. I kept some of my thoughts to myself, but as I left him I was thinking deeply. Lots of things were falling into place, in ways that were clear, but very undesirable. It did seem that the Federal government, for reasons of its own, was doing its utmost to slow down the project and increase its costs. They wanted to embarrass the oil companies in every way they could, especially financially.

There was more, and more, and more. I talked to yet another executive with Atlantic Richfield, and some of the things he told me were equally as startling.

It was about this time I noticed some unusual visitors. Who were all those men coming into Prudhoe Bay? Why, all of a sudden, are men coming in dressed in Arab garb—why are these Arabs here? What are the bankers from New York doing here? I had seen them from time to time during the two years, but now they were all converging at one time onto Prudhoe Bay, with instructions to be allowed to see everything. I knew the oil company official who had been designated to be their host. I knew him personally. Day after day he was coming to me saying, "Chaplain, you'll never guess who came through today. Chaplain, do you want to rub shoulders with one of the richest men in the world? Chaplain, why don't you ride in the back seat today? I have with me the Secretary-Treasurer of such and such a company . . . Chaplain, would you like to witness spiritually to one of the top men you'd never touch, because he would probably never go to one of your church services? . . . Chaplain today I've been designated to take a man all around through the Bay who has come here all the way from Saudi Arabia. In fact, he's coming in his own hired jet..."

Day after day, I heard talk like this, and I watched as a stream of these financial experts came to Prudhoe Bay. Why were they here? What were they coming in for? Why all of a sudden this interest in Prudhoe Bay? The money men of the world were coming from everywhere. Something intentional was going on. Something that without a doubt was planned, and now it was adding up more and more. I could see it very plainly. The pieces were indeed fitting together.

Chapter 14

The Plan to Nationalize the Oil Companies

It was 1976. I well remember that day when I walked into the office of Mr. X, and I remarked, "Sir, I sure have been having a good time lately rubbing shoulders with rich people. There's no need for me to travel around the world . . . I can meet them all right here in Prudhoe Bay. I'm the only Chaplain around," and I chuckled, "I'm the only Chaplain that can tell people that are Moslems that Jesus Christ loved them and died for them. It's been a real privilege to tell these people that Christ died for sinners—whether they come from Moslem countries, from the lower 48, or anywhere else. It's been interesting to tell them the Christian Gospel. They would not come to my church service, but they still heard the fact that salvation is available to each of them individually, if they will accept the Savior whom I love and serve."

Mr. X looked at me, interested, and perhaps a little surprised that I was able to present the Gospel in that way. However, he knew me, and had come to respect me. He knew it would be quite impossible for me to meet people and *not* give them the "Good News" if there was half an opportunity to do so.

Mr. X himself was involved in a wonderful work—the provision of oil to a needy world. I was involved in an even more important mission—to tell of the Light of the World Who had come, to tell how the Old Testament Scriptures had foretold His death, to relate the wonderful news that despite the wickedness of man, God's plan of salvation had been wonderfully foretold. And, of course, it was my joy to tell them that I personally knew forgiveness of sins, peace with God, enjoyment of the best life, because I knew the reality of walking with the risen Christ.

I told Mr. X that it had been my privilege to tell those bankers from various parts of the world that for me to live was Christ, "to die was gain," as the apostle Paul put it. I suppose those businessmen simply tolerated my point of view, but it was a real privilege to notice that they accepted me and respected my point of view. Sometimes they even listened very seriously to the wonderful news I had for them. After all, the gospel of Jesus Christ is the greatest news ever given to man.

I remember that Mr. X kind of laughed as he listened to me, and then he commented, "Well, Chaplain, where else could you get an audience like that—where else could you go in all the world to get people to listen to the gospel message in the way you presented it to these men?"

I said to him, "Sir, thank you for making it all possible. I really appreciate you letting me rub shoulders with these guys." Then I said to him, "By the way, Mr. X, why is it that all these men are up here at Prudhoe Bay all of a sudden? I used to see men like them now and then—they came through periodically, but lately we've had a flood of the biggest men in the world as far as financing is concerned."

Mr. X got up from his desk and at first was somewhat cautious. The smile disappeared from his face, and it was replaced by a frown. He closed his office door, then with a very sad look on his face, he said, "Chaplain, Atlantic Richfield has just completed the transaction of borrowing the worth of the company."

I exclaimed, "That's bankruptcy!"

He did not like the word "bankruptcy" but he remarked in his own way that was just about the size of what was happening. I had commented that it was financial suicide, and he acknowledged that was what was taking place.

At that point Mr. X and I talked again about the conversation he had with Senatore Chance back in 1975, when Mr. X had remarked that the government wanted to nationalize the oil companies.

As we carried on our conversation that day in 1976, I said to Mr. X, "You have just completed the borrowing of the worth of the company?"

"Yes, Chaplain," he answered.

I looked at him and said, "But why?"

He said to me, "Chaplain, we are struggling for survival."

I answered, "But, sir, that is not what they tell us. They say that the oil companies are huge monsters that are robbing the people of America. As American people, we have been told that the oil companies have no need of money—that they are great wealthy barons that have more than they could ever dare dream of. Why this big struggle for survival?"

Mr. X remarked, "Chaplain, the only reason we borrowed the worth of the company was that we might complete the Trans-Alaska Oil Pipeline—and in so doing, remain solvent by the sale of the oil."

Then so many things came together in
my mind. Cost overruns had caused the
costs to be increased from an estimated
$600,000,000 in 1971 to the actual cost of
the Pipeline being $12,000,000,000.   No
company could stand such cost increases
in just a few short years—and that applies
to even the wealthy oil barons. So now
Atlantic Richfield was in debt for the
amount of their total corporate worth.

At this point even more things began to
add up.   Not only were there extreme
cost overruns, but there were the false
claims of faulty welding, withdrawals of
permits, orders to dig up pipes.   There
were Union "wobblings" or "slow down"
factors. "Stop the flow of oil" seemed to
be the constant intent.   There was the
building of those $10,000 outhouses, a flo-
tila was frozen and allowed to be crushed
by the ice, and then there were those falcons—at a million dollars each!   There
were also the absurd rules concerning the
"precious tundra," and ridiculous Federal
laws and regulations, and excessive and
unwarranted fines, and more . . . .

Yes, it all added up now. Stop the oil!
And now, one of the major oil companies
of America had borrowed the worth of
the company—just to survive.   But the
American people—surely they would be
told all this? Why not ease up the restrictions, for after all there is an energy

crisis, even if it was artificially induced,
causing gas prices to go higher and higher
all the time.

Then there is the matter of interest on
$12 billion dollars. Can you imagine
what the interest would be on $12 billion
dollars? . . . at today's interest rates that
are going up all the time? This is a
struggle for survival by free enterprise.

I kind of laughed within myself as I
remembered that picture on the wall of
one of the dorms one day. It was a picture of a little child sitting on a pottie.
Beside the child was a roll of toilet paper.
As the child reached for a piece of toilet
paper, the caption under the picture read,
"The job isn't finished until the paper
work is done."

Yes, there were literally rooms filled
with paperwork. Companies had been
hired to do nothing but manage the paperwork of records. Daily, airplanes were
traveling back and forth from camps to
Fairbanks and Anchorage, doing nothing
but carrying men who were traveling to
take care of paperwork. Almost daily
some official on the Pipeline would come
to me and say, "Chaplain, I'm so frustrated I hardly know where to turn, because we've been applying for that permit
for weeks. They know the job has to be
stopped until that permit is given. All
this time my men are sitting there, doing

nothing while we're waiting on the State
to make surveys, and to decide a simple
question of a minor permit that prior to
this we had no problem whatsoever getting. In these last six months of the
Pipeline these permits are taking longer
and longer, going through the maze of
bureaucracy. The paperwork has gotten
to the point that it's momentous." It was
indeed a struggle for survival. Yes, no
doubt, the job isn't finished until the paperwork is done. But let me return to
my conversation with Mr. X. I asked
myself a question, which I then put to
Mr. X: "Sir, does the United States government own the oil companies?"

I do not remember his exact words, but
paraphrased it was something like this,
"No. The United States government does
not own the oil companies literally, but
they might as well. After all, it's their
land that we produce the oil from, on the
North Slope of Alaska, and they might
just as well have built the Trans-Alaska
Pipeline—after all, we can do nothing at
all without their permits. Not even to the
building of a section of a haul road, or
laying of a gravel pad, or the drilling of
a well, or the production of so many
barrels of oil a day from that well. In
fact, we are told almost everything we are
to do. We don't really run this job."

Does the Federal government own the

oil companies of America? They tell them how many dollars they have to spend to put a smog protection device on their refineries. They tell them exactly how the ships have to be built to haul the oil to California, and to Washington, and to Oregon, after it has been taken out of the North Slope of Alaska. If all that's not enough, they even tell them the kind of paperwork they have to turn in to prove everything they are doing.

After I put my question to him about the Federal government's owning the oil companies, Mr. X said to me very sincerely, "Chaplain, they will soon. The fact is that if we don't flow that oil in time, we will go into bankruptcy."

For the first time, I had heard it with my own ears. That was it—that was really what they were after. I finally had the last piece to the puzzle, and at last the whole picture fitted together.

I heard one of the men say one day, "I work for the purpose of paying taxes." That was it. The Federal government was aiming at total control. They knew that if they could stop the flow of oil, they would bankrupt the oil companies, and there would automatically be nationalization of the oil industry.

From this time on I looked even more carefully. I would talk to the men after I preached, and I realized that the whole

idea, for that period of six months, was
to stop the flow of oil.

At that point I had to go back and see
Mr. X again, and I did. I remembered
the day that I asked him the question,
"Mr. X, is the Federal government at-
tempting to delay the flow of oil on the
Alaska pipeline?"

He emphatically said, "Yes!" He was
angry and did not say it in a way that I
would put in this book. I would not put
in this book anything that I was told not
to tell, but that day he was very disturbed
and did not tell me or ask me not to put
it in this form. He said, "Yes, they're
trying to delay the flow of oil." Then he
continued, "I'm going to go a step further. Chaplain, if they delay the flow of
oil for a period of six months, the oil
companies of America will be thrown into
bankruptcy." He had already referred to
the possibility of bankruptcy, but now it
seemed a much closer possibility.

Then I went out to the job again. I
had heard Mr. X say it. It really was a
deliberate scheme, and I had seen it.
More and more regulations. Rules.
Withdrawing of permits—so it had gone,
on and on and on. As I talked to the
men, they indicated the same situation.
They were agreed that there was a deliberate intention to delay the flow of oil.

I went back to my room and, if I remember the day correctly, I prayed about it all that day. This is what I came up with in the conclusion of my own mind: "There is an energy crisis in America, artificially induced, and if not, why did they close down that cross-country pipeline in Wheatland, Wyoming? (We have mentioned that in an earlier chapter.) They are trying to produce an oil crisis, and if this oil was allowed to flow on time, it would produce two million barrels of oil a day, at its maximum output. That is a great amount of the precious oil that America needs."

We have said that bankruptcy would lead to nationalization. If the government managed to nationalize the oil companies, that would virtually have broken the back of private enterprise in this country.

I began to get in touch with the men even more. I made it a point to ride the line each day, to get up earlier than I had been doing, to find an oil official who would allow me to ride around with him in his truck all the day, just for the sake of talking.

As I rode with one oil company official today, and another tomorrow, and another the next day, I would keep asking questions. I was after the truth, and one oil company official would not know what the other had told me.

One day I rode with one of the men

who was responsible for certain parts of
the procedures associated with the final
flow of the oil—I will not identify him
any more than that (or the places we rode
in his pickup truck) for I want to protect
his anonymity. But I watched, and I saw
something that I could hardly believe, because I had never seen this before. I will
not elaborate the particular incident, for
it might identify the man involved. The
point is, that incident impressed on me
that there was suddenly a dramatic change
in the attitude of oil company officials.
They had "come out fighting."

By now there were two to three months
until oil flow. I had watched as the project became of immense size, and the number of men on the Slope grew day by
day, with the camps all full and the job
running full speed ahead. I had seen both
the Federal and the State governments
withdraw different permits, and literally
back the oil companies into a corner
where they had to fight.

I remember as a boy back on the farm
in Georgia, if you ever backed an animal
into a corner, even if he was an animal
that knew he could not win—if you got
him pinned up, he would fight. In those
circumstances even a small animal would
attack you in an effort to get away. That
was exactly what was happening now with
the oil companies. The government had

backed them into a corner. Time was of
the essence, for cost overruns had gone to
such a state that interest alone would eat
them up. So there was no choice—*that
oil had to flow, and it had to flow on
time!* The only way that the oil companies could survive was to flow that oil on
the given date.

How did they do it? I'll tell you how
they did it—the oil companies themselves
can never tell you the story, so I will. By
now the job had grown to such an enormous size that there weren't enough State
and Federal inspectors to keep up with
every aspect of the job. I watched, in
that last six months of the Pipeline project, as the oil companies literally bull-
dogged—if I might coin that expression—
they literally went . forward, disregarding
the stringent restrictions that had been
placed on them by the Federal and State
governments. When they were caught,
they would pay the fines, and the fines
for petty offenses ran into many thousands of dollars—but most of the time
they were not caught.

I could name incident after incident,
but if I did, it would be possible to identify the oil company officials involved, and
I do not want that to be one result of
this book being presented to the American
people. Indeed, some of those officials
might in turn be prosecuted. For that

reason I will not record for publication
specific dates. There were specific incidents on specific dates, on many occasions
when the oil companies moved forward,
disregarding the outlandish rules that had
been imposed on them by the Federal
government. Those impositions were contrary to the rules that had been agreed to
when they first contracted for this project
for the Trans-Alaska Pipeline. They literally rushed madly forward in an attempt
to survive and to flow that oil on time,
regardless of what it took to do it.

They knew the welds on the big pipeline were not faulty, they knew that the
tremendous increases in cost overruns had
been caused by exorbitant inflation and
unnecessary regimentation. They knew
that the practice of withdrawing permits
and the issuing of new permits was not
correct, either morally or legally. They
literally overran the restrictions imposed by
the government, and there was nothing the
government officials could do about it,
because they simply could not keep up
with the fast pace.

That oil was going to flow on time. I
had never seen this attitude before. Such
an attitude had not been there during the
first two years and three months of the
construction of the oil pipeline, because
during that time all regulations were very
stringently followed. All permits were

carefully obeyed, but now it was quite to the contrary.

This great animal of private enterprise had been backed into a corner, and it was fighting for survival. That was after Atlantic Richfield had borrowed the worth of the company. I do not know what the other companies on the pipeline did, but I do know what this one did, and it was the major producer on the east side of the oil field on that one 100-square-mile area from which they had been allowed to produce. So now I watched them as they literally fought for survival. They defied the government officials, because they knew it was only a matter of months and then there would be nothing more they could do about it.

I personally say at this point, "Congratulations to the oil companies." They flowed the oil on time despite a direct attempt by the Federal and State governments to stop that flow from going. It was an intentional plan to bankrupt the oil companies of America so that the oil industry could be nationalized—but they did not succeed.

I think you will agree that the incidents we have recorded make it clear that there was a very serious, concerted attempt to frustrate the oil companies and to make their costs so exorbitant that they would be forced into bankruptcy. It also seems that ultimately one of the ideas was to so discredit the oil companies in the minds of the public that they would be all too ready to allow the whole of the oil industry to be nationalized. The oil companies were to be blamed for the price of gas going up—they were to be the scapegoats, made to appear to be raking in exorbitant profits, while in fact they were being brought to the point where they were enduring a tremendous fight for survival.

Chapter 15

Waiting for a Huge New Oil Field

It was a pleasant day, with the sun shining brightly. There were very few clouds in the sky out on the Arctic Ocean —where the clouds at times looked like great waves in the sky. I woke early that morning as I had been doing often lately, to make sure that I arrived at the office of one of the company officials in order to catch a ride with him all day long. The fact was that this story was getting more exciting by the day.

So on this beautiful day of sunshine, with only a few clouds in the sky, I felt good. I went through the chow line and picked up a meal fit for a king. As I have said, that's the way the meals always were on the Pipeline—I've never eaten such good food in all my life. I think we had the choice chefs of the world to provide it.

I finished my meal that day with an expectancy of excitement in my heart. I was looking forward to finding out some new source of exuberating information as to what was really taking place in all of this planned manipulation. I put on my heavy down winter coat and my Arctic shoes, stuffed my gloves into my picket, put on my stocking cap, and laid my down cap on the seat beside me in the pickup truck. I remember how the engine ground to a start that day, for it had been cold all night. However, the engine had been plugged into an electric outlet to keep it warm and soon it warmed up and I was able to make it start. So I set off across the North Slope of Alaska for another day of excitement. What I didn't know was just how exciting that day would really be, for unbeknown to me, that day was to turn out to be one of the most revealing experiences I was to have while I was Chaplain on the Trans-Alaska Oil Pipeline.

I am quite sure that the oil company official with whom I was to get a ride did not know just what it was he was going to take me to see, because none of us really knew. You see, until after a well comes in and it is proven (*proven* is a

method they have of determining the
quantity and quality of an oil find), nobody really knows what is there.

So that morning I pulled up in front of
the building at Atlantic Richfield and
walked inside (and you will remember that
this company was responsible for building
the entire *east* side of the oil field). I
shall never forget what that door was like
on the front of their building. Have you
ever seen the doors on a commercial
freezer locker establishment? It has a
large handle on the outside and a pusher
on the inside, and the door itself is many
inches thick. That was exactly what the
door was like on the front of ARCO—it
was nothing but a big freezer door—in
reverse, of course. Every time I walked
out, it kind of reminded me that I was
walking out into a big freezer. That
freezer was called the North Slope of
Alaska which with a chill factor, has gone
as low as -130°.

Inside it was nice and cozy. I walked
up to the desk of the security guard and
asked him who happened to be in the office at that time. Usually this is what I
would do in the morning if I wanted to
have an exciting ride—I would find out
who happened to be in the office, and
then select the most likely candidate I
could and hitch a ride with him. After
all, my job as Chaplain was to be out

where the men were. So I would drive up and down the line and talk to the men while the company officials were carrying out their business. Perhaps I could do some counselling with a man who had previously come to me with a problem, while at the same time riding around on the job. In that way, I was doing two things at once.

I liked to get up on one of those big 'dozers, or get up into one of those big cranes, or stand and chat with a man while he was waiting for his buddy to finish welding a section of pipe. As I was riding around, if someone simply said, "Hi, Chaplain," it was a contact. That was part of the reason I was there. My purpose, primarily, was to help those men spiritually, and this other interest in the government's intention was secondary, *but very important, nevertheless.*

So almost every day I would ride over to ARCO, as I had done this morning. Usually the security guard would tell me of half a dozen officers, and I would have a wide choice of riding companions. One day I would ride with the equipment man, another day with Mr. X, and another day I might ride with an inspector . . . they were always quite interesting, but most of them did not want too much to do with me personally. They knew on short order that I was a conservative, and

I usually did not kow-tow to their ideas of control. However, that day the security officer named several men, and I immediately recognized one that I thought would be interesting to ride with. So I said, "Well, is he in his office or out in his vehicle?"

The guard answered, "Well, he happens to be up in his office. Why don't you just go on in. I'm sure he won't mind." So I took the liberty of going on down to the office complex and into the office of this certain ARCO executive.

He looked up as I came in, and all across his face was an air of expectancy, though at first I did not take much notice. That is usually the way these oil executives look when they see dollar signs turning over with the oil business. I looked at him with a kind of a smile on my face—I was feeling good with that beautiful sunshine outside which we didn't see all the time on the Arctic Ocean. I said, "Hey, what do you have up today?"

"Ah," he said, "You came along at just the right time. How would you like to watch something exciting? It's something that I think will turn out to be phenomenal."

"Well," I answered, "I'm always ready for excitement. If there's anything I enjoy, it's getting into something." (Of course, that's nothing new—ever since I

had been a child, if I could find something to get into . . . I just couldn't seem
to pass up the opportunity.) So I said,
"Sure, what can we get into today?"

With something that was almost laughter in his voice, he said, "Chaplain, come
on, let's ride out to the Arctic Ocean, and
I'll show you what we're going to get into
today." I could tell from the tone in his
voice that I was in for something spectacular.

"Well," I said, "Great, let's go. I'm
ready for a ride. We have all morning,
and if you like I can take all afternoon
with you, as well—that is, if it really gets
that good."

He answered, "This one *is* going to be
good."

I asked, "What do you mean?"

He just replied, "Come on, let's go."

We walked all the way down the hallway of that office complex, past the security guard and my guide told him, "If
you want me, I'll be out at such and such
a point, in such and such a vehicle."

We checked out and walked out the
freezer locker door *(into* the freezer), and
soon we had hopped inside his vehicle and
were driving west, for maybe four or five
miles. Then he turned toward the north,
and now he asked, "Chaplain, have you
ever been out to the new dock—the dock
at Prudhoe Bay?"

"Yes," I answered, "I have taken the liberty to drive up there a time or two, just to see what it is like."

"Well," he answered, "That's where we're going."

There were two docks at Prudhoe Bay. They would dock the flotilla of boats that came in the summer time—one was the original dock which had been built over by Surfcoat Camp, and that dock extended only a short way out into the Arctic Ocean. The ocean at that point was only a few feet deep. In order to bring in the larger barges that were in the flotilla during the last two years of the construction phase of the oil field, they had to go out into deeper water. After much wrangling and many battles, the oil companies were finally able to persuade the government to permit them to build a gravel pad, exactly like the gravel on the shore of the Arctic Ocean. It was a gravel pad out into the water, some two miles or thereabouts.

It was just large enough for one of those track vehicles to travel on—the vehicles that bring the flow stations, the pump stations, and injection plants after they had been brought in on the flotilla. They had huge things that I liked to call "creepy crawlers," and the tracked vehicles would carry those big buildings when they wanted to unload them from the barges. So we rode out on that gravel bar extending into the Arctic Ocean.

As we rode out to the end of the gravel road, we actually rode into the ocean. At the end of the road was a large gravel pad that extended out east and west, and on that pad they would store equipment. I remember that they had literally cut huge chunks out of the ice, for some particular purpose I can't recall. Those huge chunks of ice were almost a wall, where they had been piled up many fee thick and many feet across in diameter. We rode to a point where we could see across those huge chunks of ice, and then this oil company official said to me, "Chaplain, you are just about to watch one of the most exciting things that we oil company men will ever see at Prudhoe Bay."

I answered, "What do you mean? We are right out here at the edge of the Arctic Ocean, and I don't see anything exciting out here. There's not even any drill rigs here. In fact, there's nothing going on at this dock—we're the only people out here."

He said, "You're right, Chaplain. But I want you to look—you'll have to strain your eyes a bit—and you'll see the drill rig on a little bitty island way out there in the Arctic Ocean. If you look close, you can see it with the naked eye, without even using these glasses."

"Oh," I said, "Yes, Gull Island." The

official looked at me . . . "Oh! so you know about Gull Island, do you?"

"Well," I answered, "Someone told me a few months ago that they had taken a drill rig out to Gull Island, and I had noticed the orange colored top on that big rig out there. It just sticks above the horizon, on the Arctic Ocean, and I've heard that they are drilling for oil on Gull Island."

He said, "Yes, Chaplain, they are. Not only that, but today we are going to have the first burn from the rig—they've completed the drilling."

A "burn"—in layman's terms—is a method of proof used when an oil field or an oil well is brought in. *I was to watch that day what is probably one of the most phenomenal bits of intelligence information that has ever been discovered since the original oil discovery at Prudhoe Bay. However, this was also to be one of the most devastating things that the government of the United States has ever done to the American people in relation to the energy crisis.*

We sat there for a few minutes, not knowing exactly when the burn would take place, and this oil company official began to explain about Gull Island. It became quite interesting. He told me what I already knew, that the oil companies had been allowed to produce from only a 100-

square-mile area of the North Slope of
Alaska, yet there are many 100-square-mile
areas of land north of the Brooks Mountains, the northern-most mountain range
of the United States.   North of these
mountains there is an area of about 160
to 180 miles that slopes gradually to sea
level at Prudhoe Bay, and then out into
the Arctic Ocean.   That is the boundary,
Just a short way from the shore, of the
limit of the 100-square-mile area that the
oil companies call Prudhoe Bay.   That is
the area from which the oil is being allowed to be produced.   At maximum
flow, that Alaska oil flow will produce
two million barrels of oil every 24 hours.

So there we were, sitting out in the
Arctic Ocean, watching a speck on the
horizon . . . a speck called Gull Island.

The ARCO official proceeded to explain
to me that Gull Island is on the very, very
edge of that 100 square miles from which
they were allowed to produce. He said to
me, "Gull Island is marginal. We have
been allowed to drill there, but we know
that any angle of drilling whatsoever to
the north would mean that it would be
out of bounds of the oil field from which
we have been given permission to produce. I guess you know, Chaplain, that
this one pool of oil right here on the
north side of Alaska from which we are
presently producing can produce oil at the

rate of two million barrels every 24 hours, for the next twenty years, without any decrease in production. Not only that, but it will produce at artesian flow for the next twenty years."

That means this is one of the richest oil fields on the earth. Then he continued, "After twenty years, we will either inject water or some other liquid into the ground in order to maintain that flow of oil, but we will not have to pump this field for over twenty years. The oil comes out of the ground at about 136°F, with 1,600 pounds of natural pressure." He then further elaborated about the rich oil fields at Prudhoe Bay and stated that they have proven there are many other pools of oil on the North Slope of Alaska. He also believed that these numerous pools of oil could be produced just as easily as the Prudhoe Bay oil field. Then he told me something else I already knew. He said, "Chaplain, there is no energy crisis. There has never been an energy crisis. There will *never be* an energy crisis; we have as much oil here as in all of Saudi Arabia. If only the oil companies of America were allowed to produce it, we would have no crisis. Oh, we've been told there's a crisis, but there isn't one."

On and on that oil company official went while we sat there and idled away the time. The heater was going full blast,

because of the cold, as we were waiting
for that momentous event when we would
see black smoke from Gull Island. That
would indicate that the burn was taking
place, and we would have proof of the
finding of oil. Then we would go back to
the main office and look at the technical
data relating to what the oil companies
had found that day at Gull Island.

There was no set time of day for this
oil burn to take place, so as we sat there
waiting and watching with hopeful expectancy as to what we might actually see, we
talked about many things. We chatted
about angle drilling, and he explained to
me that they would drill an oil field
oftentimes, and after they had gone down
so many feet into the ground they would
angle off, and sometimes go many miles
at an angle. This meant that they could
drill many different wells from one gravel
pad. After they drilled those wells, they
would call them "Christmas trees," because that is exactly what they looked like
above the ground.

He explained that on Gull Island they
were drilling straight down because if they
drilled at an angle they would be out of
bounds of that small area from which the
government had allowed them to produce.
He then said, "What we find today will
prove what is on the outerskirts of this
oil field."

Then it happened!   I remember he stopped his conversation very abruptly and picked up his field glasses from beside him on the seat of the truck, and exclaimed, "Look, Chaplain! There it is!"

We both stepped out of the truck, even though it was so very cold outside—I have forgotten whether we even closed the door or not, but both of us were excited. So we looked, straining our eyes to see across to Gull Island over the ocean. They called it Gull Island because the only thing ever known to be on it was a flock of seagulls in the summer time. And there it was; a great cloud of black smoke was going up. It was almost as though a great black bomb had exploded, and the cloud grew bigger and bigger. The wind picked up the trail of the smoke and threw it to the north, and there it lay. It was like a great big cylinder churning out across the ocean.

This surely was an exciting find; there could be no more nonsense about an energy crisis now . . . *surely, there couldn't!* But I was wrong—so very wrong.

Chapter 16

Gull Island Will Blow Your Mind!

As the wind took that huge black cloud farther and farther north, it burned fiercely and seemed to turn an even deeper black. The ARCO official seemed to have an excitement about him that I had never seen before. He was elated and could hardly contain himself. He did not usually get this way . . . it was not his nature.

"This must be a big one!" he exclaimed. "Something exciting must be happening. Maybe it's another big discovery." He looked and watched, and kept looking—he stood there as though he was frozen, but he was too exuberant to freeze. It seemed as though our hands were numb because we simply could not stop watching the size of that big burn, nor could we stop the excitement caused by what we were looking at. At last he looked back at me and said, "Chaplain, I think we have just proven something phenomenal—something we have been looking for for a long time. Come on, quick! Let's go back to base and look at the technical data. Let's see what we can find out about statistics. Chaplain, I think this is going to be exciting!'' (Was that ever an understatement!?!)

We got back into the pickup truck, and he started off very quickly. He really drove fast that day. As he did so, he explained to me how you can tell what an oil well is going to produce by the burn, what the volume and the quantity are going to be, and what the pressure and the depth will be. He explained much of the technical detail as to how they drilled that well. He himself had followed it very closely, because they thought that possibly it might produce another pool of oil. They had hoped it might prove to be a pool as big as the one from which they were producing at Prudhoe Bay. If they could find another pool of oil and prove it, it would be one of the greatest finds in years.

So we rode very quickly back to the base and walked into the office. He did not hesitate for one moment to show me what it was that had been proven. He took out the statistics and showed me the papers, and let me see the proof of the find. He went from place to place that day with excitement in his voice as he told a few officials to come and look. The three or four officials that he had called gathered around to see what had happened at Gull Island.

All the time I was trying my best to find out what it was in specifics, because after all, I did not know all those terms he was using. I was a layman, and as a layman and a Chaplain, I didn't understand some of the data they were discussing, so I cannot present it here. They were so busy and excited themselves that they did not have the time to explain technicalities to me. However, I could tell by the excitement they were showing, and the way they were expressing themselves, that something big had happened.

After everyone had left the office, that oil company official said to me, "Chaplain, we have just discovered and proven another pool of oil as big and maybe even bigger than the Prudhoe Bay Field. This is phenomenal beyond words." He again said, "There is no energy crisis. Now we can build a second pipeline—now we can produce not only 2 million barrels of oil every 24 hours, but we can produce 4 million barrels of oil every 24 hours. Chaplain, this is what we as oil company officials have been waiting for."

Then suddenly the excitement was wiped off his face as he looked back at me and said, "I hope the Federal government doesn't pose any difficulty over this because of the fact that it's located on the very edge of the designated area from which we can produce." Then he looked back again and said, "Chaplain, if this is allowed to be produced, we can build another pipeline, and in another year's time we can flood America with oil— Alaskan oil, our own oil, and we won't have to worry about the Arabs. We won't be dependent on any nation on earth. Chaplain, if there are two pools of oil here this big, there are many, many dozens of pools of oil all over this North Slope of Alaska." He went on to say, "Chaplain, America has just become energy independent." I must repeat that . . . this high official of ARCO said, "America has just become energy independent."

I do not think that I have ever seen a man so excited as that man was that day, as he explained to me about that find at Gull Island.

That day I went on my way rejoicing. My, I was happy! This meant that if we could produce from the entire North Slope of Alaska, America would be oil independent! Four million barrels of oil every 24 hours—just from two of the many pools of oil! We don't have to depend on anybody. The energy crisis had just come to a screeching halt—this ought to hit the front page in every newspaper across America! This was the most exciting thing since the original find at Prudhoe Bay. Homes won't go cold anymore. American citizens will not be waiting in line for crude oil or gasoline any longer.

I think that night I hardly slept, for I had just witnessed one of the most spectacular events since the original find at Prudhoe Bay. I remember that evening as I lay in bed, trying to count sheep and trying to find some way to go to sleep. I kept going over all the things I had seen, and what I had been told. In my mind I kept trying to think about that technical data and to visualize it, and to understand some of the statistics I had seen. I thought that I would wake up the next morning and hear the entire nation of America literally shouting for joy. I thought that no longer would there be any talk of an energy crisis. Yes, we are energy independent!

Somehow in the early hours of the morning I must have drifted off to sleep, with visions of oil burns in my mind instead of sugar plums. "This means the end of the energy crisis for America" kept going through my head—for now they had proven two major pools on the North Slope of Alaska, and this oil official was exactly right and the other soundings were probably right too, and there would be many pools of oil here.

The only thing they had to do at this point was to let private enterprise loose. Let them do what American private enterprise can do so gloriously—let them do what American enterprise has done so gloriously throughout all these years. Just let them have an incentive, and with an incentive like this, gas prices would come down, so that industry could run full speed ahead. The trucks would not be left without diesel fuel. There would be plenty of gas for my vehicle! Prices?— Ha!—Tell the Arabs they can have their old oil! We don't need it. American enterprise has again done what they have always been able to do . . . they have produced. Once again Yankee ingenuity has come to the aid of the American people.

So that night I went to bed dreaming of the glory of our great nation, as a red-blooded American, proud of the fact that the Yankees had produced again, just as they always have. Yes, I went to bed on a happy note that night.

When I woke up the next morning, it was snowing outside. I had to get through the chow line right quick. I wanted to eat my breakfast in a hurry, to get back there to that camp again. I was quite sure that my excitement was shared by everyone by now, and that by the time I arrived there, the place would be crawling with reporters gathering all the data, for after all, a discovery of this magnitude should be spread all over the country.

I kind of wished that I'd called up that radio station that had asked me to give any special information, for this, of course, was phenomenal, the most phenomenal thing I had ever known. I wished I had taken them up on that toll-free call they had asked me to make when there was something special happening—oh, how I wished that I had called John Davis before and told him of this tremendous find. John Davis was with radio station KSRM, and I should have called him so he could announce this wonderful news to the whole world. I wished I had told him that they had just discovered a pool of oil as big or bigger than the one at Prudhoe Bay, so he could put it on the national wire service. I wished I had done that the night before. *Just a few hours later, oh, how I wished that!* What I would have given today, if only I had done that *yesterday!* But I didn't. The fact is, at that time, I don't think the full magnitude of that find had fully registered on me yet.

That morning I finished breakfast quickly. I remember I got in that pickup truck and cranked it up and headed off to the base camp. I didn't even wait for the truck to warm up. This was exciting.

This was phenomenal. The American people ought to rejoice over this!

I walked into the base camp, and there was nothing exceptional going on. I went by the security guard, and he was just nonchalantly sitting there, as if nothing special had taken place. I said, "Sir, where is Mr. So and So?" He said, "He's out riding around in his vehicle." I asked, "Can you call him on the radio?" He answered, "Sure."

He called him on the radio and said, "Chaplain Williams is here to see you." The man called back with what seemed to be an air of fear in his voice, and he said, "Chaplain Williams? Yes, please tell him to stay right there and not leave. I need to see him. Tell him to please wait for me in my office. I'll be in immediately."

I went to his office and sat down, and wondered why it was that on this day the trumpets were not sounding. This was a phenomenal thing, and yet there seemed to be no fuss at all about it. Sure enough, without delay, the oil company official soon walked into his office and closed the door behind him. He looked at me with a frown on his face and said, "Chaplain, what you saw yesterday, don't you ever as long as you live, let anything out that would tell anyone the data that you saw on those technical sheets."

I said, "But sir, that's going to end the energy crisis in America!"

He said, "No, Chaplain, it's not. Quite to the contrary." As he sat down behind his desk, I noticed that he was very worried, and then he continued, "Chaplain, you weren't supposed to see what I showed you yesterday. I'm sorry I let you go with me out there to watch that burn. I'm even more perturbed that I let you look at the technical data, because, Chaplain, you and I might both be in trouble if you ever tell the story of Gull Island."

I should stop at this point and state that he did not tell me not to tell the story of Gull Island, but he merely said, "We both may be in trouble if you ever tell the story of Gull Island." I watched with my own eyes what I never thought I could see in the United States of America —maybe in socialist Russia, yet—maybe under a dictatorship, but in America? No! After all, this was the country "of the people, by the people, *for* the people." Within a few days after the find and the proof of the find (proof of a vast amount of oil), I listened as that official told me that the government had ordered the oil company to seal the documents, withdraw the rig, cap the well, and not release the information about the Gull Island find. That oil field is partially under the area that the oil companies were not allowed to produce from—it is in the Arctic Ocean and microorganisms of that area might be destroyed if an oil spill ever happens. Seal the documents, withdraw the rig, and cap the well!

This company official said to me, "Chaplain, that great pool of oil is probably as big as the Prudhoe oil field, it has been proven, drilled into, and tested—we know what is there and we know the amount that is there, but government has ordered us not to produce that well, or reveal any information as to what is at Gull Island."

I could hardly believe what I heard that day. I walked out of the oil company official's office very perturbed, because again we could be lied to, the American people would be deceived again—the truth would not be told. As I walked out of that office I realized that I was only one of about six men alive who would even know the truth about Gull Island, or would ever even see the technical data. I was astonished that day because of this restriction on releasing data about the production from beneath a small island out in the Arctic Ocean. This could end the oil crisis, but I had come to the conclusion in my mind, with no doubt whatsoever, that the Federal government would never want that oil produced.

It was not the oil companies that ordered the rig removed and the well capped. It was not the oil companies that said, "We cannot go beyond our 100-mile boundary." It was not the oil companies that said, "We will not tell the American people the truth." Rather, it was your Federal and State government . . . and my Federal and State government—the officials elected by *us* to represent us for *our* welfare.

Gull Island was capped and the rig was removed, and the truth has never been told . . . *until now!*

Chapter 17

If Gull Island Didn't Blow Your Mind—This Will!

Gull Island just proved what the oil companies have believed for some time. It authenticated the seismographic findings. Seismographic testing has indicated that there is as much crude oil on the North Slope of Alaska as in Saudi Arabia. Since the Gull Island find proved to be seismographically correct, then the other testings are correct also. There are many hundreds of square miles of oil under the North Slope of Alaska.

To clarify what I am about to say, let me first re-emphasize that the government permitted the oil companies to drill and prove *many* sites (subsequently making them cap the wells and keep secret the proof of the finds), but they do not allow

them to *produce* from the wells. This is
why I have referred (below) to a *number*
of wells having been drilled (after I left
the North Slope). The *only* production
permitted is from the small area of the
North Slope.

Gull Island is located five miles off
shore from Prudhoe Bay. It is in the
Beaufort Sea.

The chemical structure of the oil at Gull
Island is different from that of the oil in
the Prudhoe Bay field and the pressure of
the field is different, proving that it is a
totally different pool of oil from that at
Prudhoe Bay.

The Gull Island burn produced 30,000
barrels of oil per day through a 3½ inch
pipe at 900 feet.

Three wells have been drilled, proven,
and capped at Gull Island. The East
Dock well also hit the Gull Island oil
pool (you can tell by the chemical structure). For forty miles to the east of Gull
Island, there has not been a single dry
hole drilled, although many wells have
been drilled. This shows the immensity of
the size of the field.

The Gull Island oil find is even larger
than the Prudhoe Bay field, which is presently producing more than two million
barrels of oil every twenty-four hours.

Where is the energy crisis? It surely is
not on the North Slope of Alaska, so it

must be only in Washington, D.C.!

Now—just in case Gull Island didn't blow your mind, try this on for size! Only recently, just west of Gull Island, the *Kuparuk* oil field has been drilled.

Again, this is a totally separate pool of oil from either the Prudhoe Bay field or the Gull Island field. The chemical make up of the field and the pressure of the field is different from the others, proving it to be a totally separate pool of oil.

In an entirely different area of the North Slope than the 100-square-mile area of the Prudhoe Bay field, the Kuparuk field is approximately 60 miles long by 30 miles wide and contains approximately the same amount of oil as the Prudhoe Bay field.

The oil in the Kuparuk field is at a 6,000-foot depth and there is 300 feet of oil sand. The field pressure is 900 lbs. at well head, and test wells have flowed at 900 barrels a day at normal flow pressure.

It is projected that 800 to 1,400 wells will be drilled into the Kuparuk field.

From 1973 through 1980 we were being told continually that America was in the midst of a major energy crisis, yet no oil production was allowed from the Kuparuk field. It wasn't until 1981 that permission was finally granted for production. *Why the delay—if there really was a crisis?*

The reason Mr. X made the statement that there is as much crude oil on the North Slope of Alaska as in all of Saudi Arabia is because the oil companies have drilled all over the North Slope and have proven there is that much oil there, but still they are only allowed to *produce* from the small area.

The North Slope is *everything* in Alaska North of the Brooks Mountains. Prudhoe Bay is a very small portion of this enormous area (just remember the *size* of Alaska, as we illustrated earlier in the book).

After the first edition of this book was printed, many people requested additional technical data. This added chapter is a result of those requests.

As I was dictating this additional material, I had the opportunity of being with a gentleman who is a speculator in oil leases. He made the statement to me, as he looked over the oath I was making public, that every oil speculator in America who is interested in Alaskan oil leases should get a copy of this, because he had never seen such pertinent information in print before. So what you have just read will excite many oil speculators and cause them to search the maps and watch for the latest leases.

Possibly you have heard it stated that the Alaskan crude oil has such a high sulphur content that it cannot be refined by most oil refineries in the U.S. We are being told that this is the reason why the Alaskan oil is not helping to solve America's energy crisis. This is also the excuse that is being used for shipping Alaskan crude oil to other countries. It has also been reported that major power companies are even telling this to their customers (in their monthly statement inserts), using it to justify their need for rate increases.

Well, here is a statistic that should silence those false claims and blow the lid off of that phony excuse of too much sulphur in the Alaskan crude.

An August 11, 1980, analysis of the Prudhoe Bay crude oil, which is flowing in the Trans-Alaska Oil Pipeline, reads as follows:

Sulphur content — 0.9%
Flash point of the oil — 35°F
Wax content — 6%
Asphalt content — 2%
Crude oil freeze temperature (better known as pour point) — 15°F

The sulphur content of the Prudhoe Bay Alaskan oil is low in comparison to oil from other sources in the U.S., as well as many foreign oils.

The Alaskan Prudhoe Bay oil can be refined by any major refinery in America without damage to the ecology.

This means, then, that the widely publicized excuse of too high a sulphur content is simply not true. Therefore, it is just one more link in the long chain of false-hoods that we are asked to believe as Americans.

An energy crisis??????

More Recent Facts— A Comparison

The following is a comparison between the three oil fields on the North Slope of Alaska which have been drilled into with numerous wells, tested, and proven.

Prudhoe Bay can produce two (2) million barrels of oil every 24 hours for 20 to 40 years at artesian pressure. Imagine what the production of the Kuparuk and Gull Island fields *could* be.

FIELD	PAY ZONE OIL	AREA OF FIELD
	(Average depth of oil pool)	
Prudhoe	600 Ft. of pay zone	100 square miles
Kuparuk	300 Ft. of pay zone	Twice the size of Prudhoe
Gull Island	1,200 Ft. of pay zone	At least four times the size of Prud-hoe . . . Estimates are that it is the richest oil field on the face of the earth.

Chapter 18

The Oil Flows— Now the Tactics Change

My two and a half years as Chaplain on the Trans-Alaska Oil Pipeline was now coming to a close. I had the distinction of being the first Chaplain assigned to the Trans-Alaska Oil Pipeline, and I had been the only Chaplain assigned to the northern sector of that pipeline, which included Prudhoe Bay oil field. When I first went there I had gone as an innocent bystander, and originally the oil companies had not even wanted a Chaplain. Through much persuasion I had obtained permission to be allowed in the work camps to help men spiritually. As stated earlier, the oil companies never paid me a salary of any kind. After two and a half years of watching, observing, hear-

ing, and seeing, I was leaving the pipeline
as a man with some definite opinions because of all that had happened.

Now it was all over—two and a half
years of many, many experiences. Gull
Island was only a matter of a few months
behind me, and the construction phase of
the pipeline was completed. Oil had
flowed on time, despite all I had seen in
the attempts to stop it. Oil was now being shipped out of Valdez into the lower
48 states, to eventually wind up in the gas
tanks of America. That four-foot pipe
was carrying a little over one million barrels of oil every 24 hours, and that oil
flow would increase with the completion
of different phases of the Valdez terminal.

Drilling at Prudhoe Bay was to continue
for quite a few years, and there was a
considerable work force left there, but up
and down that 800-mile stretch where the
men had been in camps, there were now
ghost towns. Many of the men went back
home, and some stayed in the State of
Alaska. Those crews who had numbered
thousands were now reduced to a few
men at each pump station, and those
pumps up and down that 800-mile line
were all computerized, operated out of a
great computer center in Valdez. The big
valves were automatically controlled, and
the pump stations were automatic—there
was only a monitoring system in each

pump station.

The oil was successfully flowing. There had been no leaks, except those caused by sabotage, and I have heard very little in the way of refuting many incidents that I had seen in the last 6 to 9 months on the construction of the pipeline. Today the caribou are migrating as usual. The geese are coming back each spring—I have watched them all consistently, for although I am no longer the Chaplain to the oil pipeline, I am still a missionary to Alaska. Alaska is my adopted home, and I have watched the geese come in by the thousands. Birds of many kinds migrate to the North Slope of Alaska, and the Arctic Ocean is the same as always.

As I looked back over those years, I thought, "Well, surely things will change. Undoubtedly after some period of time there will be someone who will tell the American people the story. The truth will be revealed. It will be known just how much oil there is on the North Slope of Alaska. Surely the natives of Alaska, and even the government, are interested in the royalties that could come from that oil." However, instead of the energy crisis being reduced and the truth being revealed, that energy crisis has gotten worse.

Then I began to hear more about the supposed reasons why the North Slope

Alaska oil could not come to the lower 48 states, and why we were not getting that gasoline in our tanks out in the Midwest. I heard rumors as to the excessive sulphur content in the Alaskan oil, and heard it more and more as I began to travel across America in those months that I was in the lower 48 on speaking engagements. Because of my associations with Alaska, people kept telling me about the high content of sulphur—that it was so bad it could not be used in the lower 48 states. Over and over again I heard that the lower 48 refineries could not produce the oil from Alaska, and we have seen in the last chapter that this propaganda is utter nonsense.

I had come to a new phase. As I said, my services were no longer a part of the pipeline as such, but now I found there was a new phase. Where previously the attempts had been made to prevent the oil from flowing, new tactics were now being used. It was too late to prevent the oil from flowing, for that was now history. Now the tactics were to mislead the public into believing that the oil itself was unsatisfactory, virtually unusable, and that the whole thing, that massive project of the Alaska oil pipeline, was what is proverbially called a "white elephant." The campaign against those terrible oil people destroying the precious tundra could no longer be continued and screamed from every newspaper, radio, and television, because time had proven that the ecology was not affected.

Those massive programs causing overruns into the billions of dollars had not ultimately prevented the flow of oil, but now there was a different campaign. Yes, I believe this nonsense we are hearing *is* part of a campaign: "The sulphur content is all wrong; we cannot refine it down here." This propaganda about the high sulphur was coming from the media right across the lower 48 states, and it was even coming from some oil companies, which was hard to believe.

Let me illustrate. I was conducting a missionary conference in Neodesha, Kansas. Neodesha has a very interesting place in the history of the United States, for it was there that the first oil gusher ever found in the U.S. took place. During the week of the conference I was staying in the home of the Texaco distributor for that area.

One day my host said to me, "Preacher, you know we have a real 'energy crisis' in this area. The farmers are worried where they will get their fuel from, and they don't know how they're going to harvest their crops, and the business people don't know where their gas is coming from. Right through this whole area there

is a serious energy crisis. It comes right
down to the businesses, the farms, and the
highways, as well. The reason I'm mentioning this to you is that I've been told
it's because the Prudhoe Bay oil that
comes down from Alaska can't be cracked
in the refineries in other states. Do you
know anything about that?"

. The word "crack" is a term that is
used to refer to the oil being broken down
into auto fuel, aviation fuel, diesel, etc.

I said to my host, "I don't know anything about that, but when I get back to
Alaska, I shall make some inquiries. It
happens that I know the man who designed the cracking plant at Prudhoe Bay,
and we should get an answer. This man
was there when the first wells were sunk,
and he is an important man at Prudhoe
Bay."

About two months later I was back in
Alaska, and again I saw Mr. X. I told
him I had been in Neodesha, Kansas, and
that while I was in the home of the Texaco distributor he had asked questions
about the problems in breaking down the
Prudhoe Bay oil in American refineries.
I mentioned that my friend had said that
the oil had such a high sulphur content
that it simply was not suited to these refineries.

I myself knew that this matter of "high
sulphur content" was a pet peeve of the

ecologists, and I was interested to see the reactions of Mr. X.

He literally laughed. I'll never forget the way his face lit up and he burst into laughter. "Is that really what the man told you?"

I said, "Yes, sir, it is."

At that point Mr. X reminded me of his own position, and of the long association he had with the oil company. It was he who originally designed and then had arranged for the building of the cracking plant at Prudhoe Bay, this being the plant that produces the fuel oil, the automobile oil, the jet oil, and the various other types of oil produced by that plant. The oil so produced is used for various purposes at Prudhoe Bay, and for the entire area to the south, as far as the Yukon River. He had been there at Prudhoe Bay at the time when the first well produced oil, and he had analyzed the samples taken out— and from all the other wells in the area. He reminded me that he was able to speak with authority and certainty on the matter of sulphur content in the Prudhoe Bay oil. Then he said to me, "*The oil at the Prudhoe Bay field is pure enough that it can be cracked by any refinery in the United States, with only minor refinery alterations. Prudhoe Bay crude oil contains only 0.9% sulphur, which is quite low.*"

I knew that was true of any refinery—
and that it was necessary to adapt the
plant to refine any oil from another area
or some other part of the world. Such
adaptations were not uncommon, because
oil comes from so many different areas.
Mr. X went on, "The sulphur content
from the Prudhoe Bay is not excessive. It
certainly is not a major problem. Alaskan
Prudhoe Bay oil can be used very readily
to supply all the needs of all the people
of the United States for many years to
come."

I thanked Mr. X, and soon went on my
way. Once again I knew that this fitted
into the overall picture (which, as you
will recall, is nationalization of the oil industry). I had seen a number of newspaper reports, and heard spoken commen-
tary on the media to the effect that Alaskan oil had too much sulphur to crack in
U.S. refineries. Once again this was
shown to be a prejudiced judgment, without basis. I might add that other oil
company officials have since confirmed the
authoritative statements made to me that
day by Mr. X.

It is relevant to notice that there have
been other press reports to the effect that
the Alaskan oil field is drying up. One
wonders whether such reports are deliberate scare-tactics, or intentional distortions
of fact. It is certainly true that huge

quantities of oil are available from Prudhoe Bay, and from other areas of the
North Slope of Alaska.

Could it be that the government of the
U.S. might not *allow* the refineries to
make these modifications?   Could this
somehow be done again under the guise
of protecting the ecology? Yes, that could
be what the next step was. So now they
were suggesting that they barter the oil,
let some other country take the Alaska
oil, and then more of other countries' oil
would come into America.   It was very
plain that this was yet another part of the
scheme to make this nation dependent
upon other nations for its supply of oil.

Then how about all the rest of that oil
at Prudhoe Bay?   How about the fact
that Mr. X had said that there was as
much crude oil on the North Slope of
Alaska as in all of Saudi Arabia?   What
about Gull Island, of all things?   Then I
watched.

I sat back as a good American citizen,
praying and hoping that someone would
properly and profitably inform those in
high positions. I remembered what I had
been told by Senator Chance when he said,
"I was in the Senate Chambers of the
State of Colorado when the men from
Washington came to talk to us as to why
there was an energy crisis, and about the
severity of the energy crisis."   After one

week on the North Slope of Alaska, Senator Chance had said to me, "Almost everything said to me by those briefers from Washington, D.C. was a lie."

Six months went by, and the oil was flowing. One year went by. I thought to myself, "No one is coming out with the truth yet." A year and a half went by, and then I saw it again. I saw it again, the same identical thing, except that this time it was disguised under a different heading.

Now it was price increases. Yes, every few days the prices at gas pumps were going up and up. They said it would reach a dollar a gallon, and we Americans said there was no chance of it ever getting that high. Then it got to a $1.50 a gallon, and now they are saying $2.00 per gallon. *WHY? There is no shortage.* There is no *genuine* oil shortage.

There is plenty of oil here. It is all over the country.

Then I began to analyze the new strategy that seemed to be coming out of somewhere, and I found there were all kinds of other regulations being insisted upon. I learned there were regulations that said that we must put filtering devices on all gas stations across the nation, so that no fuel fumes escape into the atmosphere from the trucks that deliver the fuel. All the fumes left in a truck have to be recycled. So at exorbitant expense again, it is being insisted that there must be special gadgets put on those trucks, and on all the vents of the filling stations all across America. They told me then that the price of fuel must go up two cents per gallon in order to pay for that. It began to be clear to me that there was another campaign on, to make the fuel companies look like fools.

Then one day Mr. X and I crossed paths again. As we did, I asked him a question. "Mr. X, now that the oil has flowed and the oil companies have remained solvent, contrary to what the Federal government seemed to want, could it possibly be that the campaign now is to make the oil companies look like fools? Are they being made to show exorbitant price increases and likewise being made to *appear* to show exorbitant profits? Is that why there are these new regulations that make the price of fuel go up and up?"

(He looked sort of stunned, as if I had been reading his mind.) He answered me, "Yes, Chaplain, there does appear to be a move on today to so disgrace the oil companies in the minds of the American people that some day the people themselves will *ask* the government to nationalize the oil companies."

Price increases. Regulations. Then I said, "Why don't you tell the truth about those price increases?''

Mr. X again remarked, just as he had to Senator Chance that day, ''Chaplain, we can't. We don't dare tell the truth. As oil companies we can't tell the entire story. After all, the Federal government has already imposed so many regulations and stipulations over us, and there are so many laws held over our heads (laws that have never yet been strictly enforced), that if we ever told the truth in its entirety, then by the enforcement of laws that have already been passed, we could be forced into bankruptcy within a year's time.''

At that point I decided it was time for somebody to tell this story, the story of a scandal greater than Watergate. Then came the Republican men's committee dinner in Denver, Colorado. Someone heard that I had information about the energy crisis, and I was asked to be the speaker. I gave the truth, and I think that was the first time I ever presented it in public to a general audience of that kind.

That day the men seemed to be fascinated, and soon there was another speaking engagement, and another, and another. It began to mushroom, and I decided it was time to put this story in print. It is necessary to do this so that everyone can know it, if they are willing to believe it. It was not just a matter of what I supposed it to be, for I have largely avoided opinions. These are the facts as I actually saw them.

Then one day, after several speaking engagements, I met a man who had a good position with one of the major oil companies of America. He came to me after I had told the facts of Prudhoe Bay, and he questioned me at length about other things beside what I had said at that meeting. However, as our conversation continued, I asked him a question, "Sir, were you in accord with everything I had to say today? And have you ever been to Prudhoe Bay?"

He answered, "I have been to Prudhoe Bay." Then he went on to say, "No, I am not fully in accord with everything you have said today."

"Why not?" I asked.

He said, "Because I do not believe that there is that much crude oil on the North Slope of Alaska."

I asked him, "Sir what makes you think there is not that much oil there?"

He answered, "I am a geologist, and I was on the North Slope, and I went to Prudhoe Bay."

I said to him, "Then, Sir, you must know about Gull Island?" He said that he had heard something about the Gull Island find. I said to him, "Then you must know that there is another pool of oil there as big as the Prudhoe Bay pool?"

His answer was, "Now, we were informed that the Gull Island find was very small and insignificant, and we were told that the proof of find there indicated it was not worth production."

We continued talking, and several other topics were discussed. He then said to me, "Chaplain, I hope you will not make drastic statements about how much oil is at Prudhoe Bay." Then I decided that I would pin him down. I asked, "You were at Prudhoe Bay." "When were you there?" He answered, "Oh, only for the first few months of production back in 1974." I persisted, "How long did you actually stay there?"

"Oh," he answered, "I did not actually stay there. I was just in and out of Prudhoe Bay periodically."

I said to him, "Sir, the Gull Island find did not take place until 1976. How could you know the details?"

"Well," he answered, "To be honest with you I really don't. I only know what I heard."

I left the meeting that day, knowing that the "powers that be" had successfully spread false reports across America, in an attempt to make the American people believe that there really is not the quantity of oil in Alaska that they originally thought was there.

But, you see, I know different, because I *was* there. I lived there for 2½ years. I was there in summer and winter. I watched the well come in. I watched the burn. I watched the proofs of find. I saw the technical data. I looked at the statistics. I saw the sheets that represented the seismographic tests and talked with the officials. I lived in the dorms. I rubbed shoulders with the oil company officials of America. I was allowed to ride about freely across that North Slope area in my own vehicle, as well as with company officials. I was allowed to see what was there for myself. Today I can declare only what I saw, just as it was. That is not always what is published today, but it is as I saw it, as it literally exists.

Another oil company official spoke to me one day where I had been speaking in another men's committee meeting. He came to me after the meeting and said, "Chaplain, I also am with one of the major oil companies." As he shook my hand he said, with a big smile on his face, "I sure am glad to see someone willing to tell it like it is."

I said to him, "Sir, why do you say that? You say that you are with one of the oil companies—*why can't you tell it as it is?*"

He said, "Chaplain, we tried, but it doesn't work. Every time, someone stops us. We cannot tell it as it is because they think we are biased. After all, we're paid by the oil companies." He then said, "As an oil company official, I just want to shake your hand today and say one thing: I concur with what you said. Congratulations! Go tell it to the American people, because we can't."

That is the intention of this book—for I believe that we are faced with a scandal greater than Watergate.

Chapter 19

The Energy Non-Crisis of Natural Gas: A Startling Prediction Comes True

Again it was 1976, and I had only a few more weeks to stay in Prudhoe Bay before leaving the construction phase of the Trans-Alaska Oil Pipeline. It was almost completed, and my work as Chaplain was virtually over. I watched the last flotilla that came to Prudhoe Bay in 1976, and I saw them bring in monstrous buildings and equipment, the size of which I had never seen before. It was equipment driven by huge turbines, such as with huge jet engines. I saw sophisticated equipment, including huge separating tanks that had been especially designed and coated inside and out for the separating of the crude oil and the natural gas as it came up from the wells.

I watched monstrous-sized burns. They were not allowed to flare and burn the natural gas as they used to do years ago with the oil, but now I watched a burn so immense that the amount of cubic feet of gas that it could burn each hour was mind-boggling. After the pipeline construction phase was over, Mr. X and I happened to be talking one day (I still lived in Alaska as a missionary and we were visiting), and I asked him what they were going to do with the natural gas that came from the Prudhoe Bay oil field.

Mr. X said, "Chaplain, do you remember all those huge pumps, and the large separating tanks, and those four-story buildings that were brought in on the flotilla of 1976? Do you remember that injection station? Do you happen to remember all those extra wells that were being dug—over by Atlantic Richfield's main complex? And do you remember that huge flare not far from there, that was flared out across the water of the Arctic Ocean, because they won't allow us to flare into the air like they used to do?"

I said, "Yes, Mr. X, I do remember all that. I saw it just before I left. In fact, one of the men took me over to the building and showed me through it. I could hardly believe the size of those huge containers inside those huge buildings they called 'separators'."

"Well," he said, "Chaplain, we've been promised that we could produce that natural gas. We've been promised that the Federal government would allow us to build a natural gas pipeline down the same corridor from Prudhoe Bay to Valdez, and there to liquefy the natural gas. That was the original proposal, and we have built the pipeline down the corridor with the intent of carrying the natural gas line, as well. It was to be taken down the corridor from Prudhoe Bay to Valdez, then liquefied and taken out by ship to the lower 48 states. Then it was to be re-gasified, and eventually sent by pipeline all across America."

I said, "Yes, Mr. X, I remember that was the proposal. In fact, there are still many hundreds of men in Alaska right now who are here for the purpose of being a part of the construction job to build that natural gas pipeline. I remember the technicians and the engineers telling me in each of the work camps that they expected the natural gas pipeline to be constructed very shortly. Most of them were expecting that project to start just as quickly as this project ended."

"Yes," Mr. X said, "That project was to start on the heels of this one, but I am predicting now, Chaplain, that the natural gas of the Prudhoe Bay oil field will never be produced."

I said, "Mr. X, don't I remember you telling me one time—about two years ago —that there was a plentiful amount of natural gas right in the Prudhoe Bay oil field alone?" ·

"Yes," Mr. X answered, "There's enough natural gas on the North Slope of Alaska to provide the entire United States with natural gas for the next two hundred years. If every other natural gas well in America were shut off, there would still be enough natural gas on the North Slope to provide for the total projected natural gas needs for all of the United States for 200 years. That is based on the present calculated rate of consumption and the expected increased consumption year by year —there's still enough there to provide all the projected needs of the United States for 200 years."

"Well," I commented, "Mr. X, we've been told there was a natural gas shortage, as well as an oil crisis."

Mr. X kind of laughed, "Chaplain, that natural gas pipeline will never be built."

I asked him, "What are you doing with the natural gas at Prudhoe Bay? That gas comes out of the ground right along with the oil. You have to do something with it."

He told me, "Yes, Chaplain, we have to do something with it. We cannot burn it —they will not allow us to. Therefore, it is costing us millions and millions of dollars to build huge facilities, and to drill additional wells and to provide huge injection pumps to pump that natural gas back into the ground. We are pumping the same gas back into the same field that it came from, at many, many cubic feet per day."

"Well," I reflected, "It would be no trouble to build a natural gas pipeline down the same corridor, and to bring the natural gas to America. After all, it is already available—the well has already been drilled, and the corridor itself is available. The pump stations and all the facilities are already here. Even the camps would soon be ready to be occupied again."

Mr. X agreed. Then he gave a startling prediction that came true only a few months later.

You who read this book should mark it clearly in your own mind, for it is very significant. He made the projection that President Carter would have a choice as to which way the natural gas pipeline would be constructed. It could be built down the same present corridor, from Prudhoe Bay to Valdez, then liquefied in Valdez and taken out by ship to the lower 48 states, or President Carter woud have the alternative of having the natural gas pipeline built across Canada. That would be over 3,000 miles to the United States (3,000 miles of pipe across another country, rather than 800 miles of pipe across an already existing corridor down one of our own states).

Mr. X said, "Chaplain, I predict that when the President comes to the time of his choice, he will choose that the natural gas pipeline must be built across Canada, and that it cannot by any circumstances be built across Alaska, liquefied, taken to the West coast, and then distributed across America."

I was confused and asked, "Why do you predict that? What is the significance of that?"

He answered, "Within 6 months time we could be producing natural gas for America, down the present corridor which has already been built for it. The natural gas could go down the same route that has been used for the Prudhoe Bay oil. A natural gas pipeline could be built down that corridor within six months, and a liquefaction plant in Valdez could be built in a few months, because everything is ready to go. Within a matter of months the natural gas could be distributed across the entire United States, with the addition of a few cross-country pipelines. If that were done, there would never be any crisis of natural gas in all the lower 48 states of America, and that could all be accomplished within a matter of a few months." He paused, "But Chaplain, if the President chooses to take the natural gas pipeline across Canada, it will never be built."

Again I pressed him for an answer: "But sir, *why* could it never be built?"

He told me, "First of all, the reason it will never be built is that with the rate of inflation in America today and around the world, there is no consortium of gas companies in the world that could afford to build that pipeline. They could not raise that kind of money. Chaplain, the red tape that is involved, and the approval by Canada, for that pipeline to be built across their country rule it out—plus, of course, the royalties that Canada would require of us. There is no way the red tape and all the other details would be completed in your lifetime or mine, to carry that pipeline across Canada."

Then I asked, "Mr. X, are you trying to tell me that President Carter and his advisors intentionally chose for that pipeline to go across Canada, because he

wanted to stop that gas from reaching the
lower 48 states?"

Mr. X looked back at me and said,
"Chaplain, that is exactly right." So that
natural gas from the Prudhoe Bay oil field
—a field that could provide the entire
United States with gas for 200 years—will
never reach the lower 48 and our big cities. The reason is that the government
has deliberately chosen that it will not
reach the people of America.

I said, "But Sir, surely the government
officials must care something for our people?"

Mr. X said, "Chaplain, mark my words
that natural gas pipeline will never be
built."

"Well," I said, "Sir, what will happen
to the Prudhoe Bay oil field? I've been
told that the natural gas top has to come
off after a certain period of time, or the
oil field cannot be properly produced, because there will be too much top pressure."

He nodded. "Yes, Chaplain, that is
exactly correct. If something is not done
within five years, we will begin to have
difficulties with the production of the
crude oil from the Prudhoe Bay field."

Then I said, "Mr. X, could this also be
possibly a part of the great plot—to somehow shorten the life of the Prudhoe Bay
oil field, because they will not allow the

natural gas to be taken off and used, or to be burned?"

He just looked at me with a rather unusual smile, as if to say, "Well, even you, Chaplain, have sense enough to know that!" I then asked, "What will you do over a period of time?"

He said, "After so many years we will have to inject water into the ground, and hope we can keep up the pressure of the field to the point where we can maintain production over the number of years that we have projected that the oil field can produce."

So what about natural gas? No, it will never reach America.

And what about Alaska? You guessed it! Morris Udahl's bill came along, so now we will take most of the land in Alaska, and lock it up in wilderness areas for all time and eternity. This was just one more part of the great plan to lock up all the energy that is so abundant in the North Slope of Alaska. The D-2 land bill has passed, the natural resources can never be produced. It can never be drilled, and it can never be used. We will never be allowed in to find out more, to make the tests to see what is there. They say it is being preserved for our future generations. Future generations? With the technology of today, you mean we cannot develop some alternative means of supplying energy? . . . even when we have at least enough (with crude oil and natural gas) to supply our nation's energy needs for generations ahead from just a few pools of oil on the North Slope of Alaska? What is the real answer? If a satisfactory alternate energy source cannot be discovered and developed in that length of time, it's because *nobody* is trying . . . *or somebody doesn't want one found!*

Not only has the crude oil been lied about, but now the natural gas has been stopped, too. You as "John Doe American Consumer" will not be allowed to burn that cheap natural gas. It could be in your home in time for the *next* winter if only private enterprise were allowed to go in and produce what they have already drilled into. Within six months they could be moving it down the same Alaskan corridor as the oil . . . as they were originally promised they could do.

Chapter 20

A Scandal Greater Than Watergate?

What kind of scandal would be greater than Watergate? Today it has appeared on the scene, and its ultimate objective is to get Americans to agree to—in fact, to *request—socialization.* Why have oil prices gone so high? I have never heard anyone reveal the true reasons before. I wondered if they would allow me to tell them . . . but it's high time someone did, so I am going to! Remember Shylock and his pound of flesh? Well, this is close! Can you imagine what the interest would be on $12 billion?—even at the best prime rates? Let's even deduct the $2 billion the the pipeline was supposed to cost once it got going. Let's just call it a nice round $10 billion in cost overruns. One company, in order to pay their proportionate share, had to borrow an amount equal to the entire net worth of the company . . .

to literally mortgage the whole company, as it were. Can you just imagine the interest on all that? Such staggering sums are mind boggling—and generally we tend to associate them with nothing smaller than the "National Debt."

Perhaps you have read statements by the oil companies telling us that their "declared" profits don't really give the true picture. Even such articles as those in *Reader's Digest* don't reveal the whole story. Oil companies are not allowed to include those exorbitant interest charges as deductions from profits, which for you and me would be the usual procedure. Remember the atrocious prime interest rates that we face today? Well, work it out! What would be the interest on $10 billion? Now, mentally deduct those staggering figures from the highly-publicized "declared" profits . . . then you tell me who is really the villain! Certainly not the oil companies!

Then there is the matter of that tax on "Windfall Profits" that the oil companies make. That is yet another ploy to weaken them. We hear a lot about windfall profits, but how much publicity has been given to their huge interest charges?—virtually none, if any at all.

By this time you are probably way ahead of me . . . in fact, I'll bet you've already guessed who is going to pay that high interest? Apart from the oil companies themselves, there is only one person who *can* pay for it today—you, me, the consumer, John Doe! The redblooded American is going to pick up the tab for that exorbitant interest—every time he drives up to the gas pump. The one who believes in free enterprise is the one who will pay for that interest—John Doe, who stands for the principles on which our forefathers founded this country, one of which is incentive, not socialism. (Or, if you prefer, you could nationalize the oil companies and pay instead for the government bureaucracy—but then we've already seen how well *that* works!)

Throughout this writing so far, I have attempted to give only actual observations, and I have deliberately withheld my personal opinions as much as possible. However, I am sure you will recognize that in Watergate it was necessary to read between the lines. Likewise, in the scandal that is greater than Watergate, you must again read between the lines. This scandal not only touches the government, but it touches every minute area of every American's life—for that's the name of the game with something as crucial as energy.

Let's do a little imaginative "supposing" now. If I were a government that seemed to have socialistic trends, and apparently wanted to control the lives of every single individual living in this nation, I would first need to control energy. An excellent way to gain this control would be to cause the American people to think that prices were going so high simply because the oil companies were making exorbitant profits (or perhaps better yet, make it look like there is a shortage because they are inefficient, etc.). Of course, you would fail to mention that not all the expenses are deducted before the "profits" are "declared" (as is customary in other business reporting), because if they (the consumers) become adequately disturbed about the "rip-off," they will probably even *demand* that these vital services be nationalized for everyone's best interests. Since it is now too late to stop the flow of oil, the scheme has switched to getting the American people to believe that the oil companies are amassing great fortunes from the North Slope oil, therefore, they shouldn't need to raise the prices.

Reading between the lines? After all, the government has never told us what the interest would be on $10 billion, not to mention the "small matter" of the repayment of the $10 billion principal itself, just to pay off the cost overruns forced on the oil companies by the Federal and State governments. I would hate to pay the bill—but I am paying it . . . *and so are you!* The prices at the gas pumps are going up, and up, and up (in spite of regional temporary declines). We will eventually become so disgruntled with the oil companies that we will actually request the Federal government to take them over and nationalize them?

So now the push is no longer to stop the flow of the oil—*it's a little late for that,* for the oil is already flowing from one of the greatest oil fields in the world. So . . . today the move is price increases—regulations—cut backs—energy control—"Conserve. The world is running out." Why is fuel approximately $1.50 a gallon (as this manuscript is being written)? I'll tell you why. It is because you have to pay the interest on $10 billion, in addition to paying back the principal. And don't forget those who have placed themselves in charge of "over"-protecting our environment, along with the many added costs they bring . . . *much of which is totally unnecessary!*

Today, as long as inflation continues its upward spiral, the Federal government makes more and more money from every John Doe American, because as your wages go up, so does the government's share. By the same token, the tax *structure* makes Uncle Sam even richer, because they are not revising the tax structure accordingly as fast as the inflation is escalating. As inflation continues to spiral (and your salary does not go up as fast as that spiral), America gets deeper and deeper into debt and ever more dependent on the rest of the world. (One of the solutions, of course, is greater [American] production at lesser cost.) At the same time, the very ones who are the champions of free enterprise (industry) are stymied because they are not allowed to produce. Industry is struggling to survive because the thumb of "big daddy" government is crushing it every day it continues to exist.

In the year 1973, we experienced the first real so-called energy crisis per se. By the way, have you ever noticed that each of these energy crises have affected only one portion of the country at a time? In 1973 it was only the East Coast (the northern part, in particular). There was no crisis in the West. There was no crisis in the Midwest. There was no crisis in the South. Why the Northeast? Because, you see, that was the first testing ground to find out how far the government could take gullible Americans. Then about the time folks were ready to revolt, suddenly there was no longer a crisis in the Northeast. All of a sudden, out of a clear blue sky, for no known reason, it ceased to exist . . . all the gas you wanted!

Next, if you remember, it was California. The lines had disappeared in the

Northeast. Then they thought, "We'll try the farming section of the country." However, that one did not get too much publicity, so that "crisis" didn't last too long.

It seemed strange to me that I was told by oil company officials a number of months in advance where the next "crisis" would occur.

One section after another of America has been tried, to see just how far they could be pushed before they rebel. Then, at the point of rebellion, the government backs off. All of a sudden there is no energy crisis in that area anymore.

How about the natural gas crisis? Do you remember everyone was screaming about it in 1974, and all across the country people were going cold in their homes because of the so-called shortage of natural gas? Then once more—somehow—out of a clear blue sky, there's plenty of natural gas Why? Because there never was a shortage—the shortage simply never existed. It was "created" for the express purpose of finding out just how far you and I could be pushed before we rebel.

Then, quite recently (only a few weeks ago as I write), something else quite startling happened. I was talking with a certain gentleman in the Midwest who lives near Estes Park in the center of Colorado (that is one of the largest parks in the

Midwest). This man said to me, "I live near Estes Park. My home is only a short distance away from that area, and I have noticed that up in the forest area of Estes Park there are some odd looking structures that are somehow being hauled in by huge helicopters, and they have been drilling in the national forest." He continued, "I wondered about that—after all this was supposed to be a closed area, but they *were drilling* and then they would move. There are a number of big drills and that rig in there—they were somewhat camouflaged so that nobody would recognize them, but since I live right near Estes Park, I could not help but realize that something was going on. I kept noticing the big helicopters moving this big equipment in. As well as that, some of the workmen lived right around me, and day after day they were driving in and out, and there was drilling for oil going on, right there in the park itself."

The man's story was becoming interesting. He went on, "You see, I am also a fire fighter, and it is my job, when a fire develops in the Estes National Park area, to go in and help them stop that fire. We have been extensively trained for working in our area, and we know every part of those mountains—and exactly how to fight a fire in them. Last Summer, sure enough (as often happens in the summer time) a fire developed in the Estes Park forest area. The fire fighters were called out by the local officials, and everybody gathered together ready to fight the fire." Then he said to me, "Chaplain Lindsey, we will never understand why, but the BLM [Bureau of Land Management] came in and said, 'cancel the fire fight. We at the BLM will take care of it. We'll handle this one ourselves.' Our local officials said, 'But you're not trained for it. You do not have the manpower. We have men trained and they're *supposed* to do it.' However, the BLM said very emphatically, 'No one will go into that forest area.' "

The man went on to say, "Chaplain Lindsey, they did not go in. And they let it burn. They attempted to contain parts of it, but parts of it they could not. It burned a vast area, but we were not allowed to go in and fight it. Afterwards, it turned out that many of the rigs had been burned, but they started all over again. They're all very secretive about that—why would they not let us know what was going on in that area?"

This is why: The man told me that he had probed very thoroughly into it and had learned that they had made a very sizable oil find. The government itself had authorized most of the drilling, but after they found it, they capped it. They said,

"It will not be produced."

This same thing can be multiplied in Wyoming and in other oil-productive areas all over the country. The companies have been ordered *not* to produce. The finds are there. They know the oil wells are there.

Recently, I talked to a certain group on this subject. Afterward, a gentleman came to me. He said, "Chaplain, it is my job to go around to the different areas of Wyoming. I check the level of the big oil tanks and the oil that is being pumped out of the ground. I've been working at this job for a number of years." He then related the following story. He named things that had happened a number of years ago, but I have withheld details and particulars that could lead to the identification of this man or his area. He said, "Some years ago we were producing X number of gallons, but in 1974 they cut back the number of actual pumping actions that our pumps make every 24 hours. That is to say, that a pump that was making X number of pumps 5 years ago, today is making only a portion of that number of pumps. They had slowed the pumps down."

I said, "Why sir? We're supposed to be in an energy crisis?"

He answered, "I've asked myself that many times. The same field used to produce X number of gallons, and it still has the same *number* of pumps and everything is working like it used to, but now they've cut back on the *pumping action* of those pumps. Today it takes many more days to pump the same number of gallons from the identical field—it is the same field they've been pumping for years."

So there appears to be an intentional cutback in the production of the oil fields of this part of Wyoming. Why?

I could go on and on and on. As I travel across America lecturing, I meet people all the time. I have many speaking engagements in churches, business and civic organizations, and that story can be multiplied and multiplied. It certainly does appear that there was a greater scandal than Watergate. Why?

I am convinced that there is a definite reason, and at this point I move from observations to personal opinion. There is only one thing on earth by which every human being can be controlled, if that product itself is controlled. That product is energy. The world today has become dependent on energy—for its homes, its lights, its fuel, its automobiles, its airplanes, its trucking industry, its railroads, its delivery of goods, etc. Electricity is produced by the energy of today. Every facet and aspect of our lives can be controlled when energy is controlled. There

is no other product on the face of the
earth that can so control the American
people—and all the people of the world.
Whoever controls the energy . . . *controls
us!*

The fact is, if energy can be controlled, *you* can be controlled. It could
not be done by money, for methods of
bartering could be developed by the people. If your *energy* is controlled, however, then "Big Brother" can control how
you live in your home; when you go and
where you go; the products you buy; the
style of life that you will live; even the
level of life at which you will live. They
can control your state of life and your
every movement.

In the days of the horse and buggy, this
would not have been so, but today we are
dependent entirely on energy. Therefore,
because of our complete dependence, we
have become ready targets. Now, if they
can brainwash the people into believing
that there is a true energy crisis, *when
there actually is not,* then they can slow
down our society, they can destroy our
free enterprise way of life, and they can
control every area of our being. It certainly points ultimately to one-world control . . . and to an evil dictatorship. Absolute power corrupts absolutely.

Where does it all begin? It all begins
with those in high places who are willing

to control you and me. *Money* is not the question in the energy crisis today (even though it may be hard to believe while we shell out so much for energy). The price of gas at the fuel pump and the price of fuel oil that goes into your home for heating (or gas or electricity)—those are not the real factors. Price is not what they're after, even though they would like you to believe that it is. The motivating force today is control, power, manipulation, the ability to regulate every area of life in such a way that you can be brought completely under the domination of the system and, in turn, those who manipulate that system.

At one time I, too, believed the oil companies were the "bad guys," until I saw the oil companies struggling for their very existence. One time I, too, thought that the government was "of the people, by the people, *for* the people."

There are a few in government who still live by that creed, and I hope that one result of writing this book will be that others will return to that point of view. (If they don't we need to let them know that after the next election they will be out job hunting!)

However, it is undoubtedly true that the great motivating concepts today are power and control (which are almost synonymous)—and surely by now you must see that this is not only in relation to energy, but it applies in other areas of life, as well.

So what is the conclusion? It is rather startling, isn't it? It is the conclusion I came to after a number of years of examining the facts and putting pieces together. We are being sold down the river and we'd better put a stop to it before it's too late. As Mr. X said to me at Prudhoe Bay some years ago, "There is no energy crisis. There has never been an energy crisis. There will never be an energy crisis, but rather it is the purpose of the 'powers that be' to *produce* an energy crisis. Because, you see, Chaplain, there is as much crude oil on the North Slope of Alaska as in all of Saudi Arabia. [Not to mention all the rest of the places they have discovered oil and just capped it off!] If free enterprise were allowed to produce that oil, America could be independent for energy within five years time."

"Not only that," Mr. X related, "The United States could be financially free of the rest of the world within five years, if only private enterprise were allowed to develop natural resources—the energy . . . crude oil and natural gas—that are found on the North Slope of Alaska today."

Do they really *want* us to be free of the rest of the world, or is it total control they are aiming for? The answer has become increasingly obvious and the pace has accelerated. Soon our current inflation rate of 18% will seem trivial. Your money will be worthless. You will, in fact, be controlled by a computer, which, in turn, is controlled by a dictator who will manipulate those under his control like robots. (The truth is that because of the *convenience* of all those sophisticated new systems, we are even now being lulled into complacency about retaining control over our own activities, in the area of banking and other financial transactions in particular, i.e., automatic deposit of your paycheck, automatic disbursements from your account, automatic grocery checkouts [by little lines on the packages], "telephone" bill-paying, automated 24-hour "teller" [a nice word for a machine], *et al.* We are being very subtly programmed to do business with machines [computers] instead of people.) The dictator will bring men and women into willing subjection by his favors, e.g., homes, finances, jobs, etc. All the time they themselves are—like flies—being drawn into his worldwide web of intrigue.

Stop it! Stand up and be counted! Soon it will be too late. John Doe Citizen, himself, must take action—*NOW!*

Chapter 21

Some Dramatic Conclusions

One of the most exciting and characterized histories of the whole world encompasses the birth of our nation. Independence Hall, the Declaration of Independence, the war against the Red Coats, and Liberty Bell—they all speak of a charactered, responsible people who paid a great price in order to give birth to a nation which was totally committed to the principles of freedom.

Our beginnings were linked with that remarkable political document, the Constitution. It is the very form and system of our government, with its built-in checks and balances, its free elections, its unparalleled Federal and State election system, and its quest for justice (with dual levels of court systems). It is unmatched in past history or anywhere on the international scene today.

The United States was born with the greatest explosion of freedom ever expressed on the pages of human history. It grew with zeal and excitement in the 1700's and passed through its adolescence to a strong military and industrial adulthood in the 1800's.

By the 1900's that freedom had reached its middle age, but during this period it grew fat on secular materialism and contracted the highly contagious, fatal disease of humanism. Today in every segment of society we are observing many evils—social disorders, strained foreign relations, and military anemia. These are all symptoms of a dying culture, and they sadly point to the death of freedom in the United States, perhaps even during this century.

Most of the nations of the world have already suffered and died from the same disease. In most countries of the world, personal freedom is already practically nonexistent. For the most part, the world has become decadent in its lifestyle, its literature, and its entertainment. The currency is nearly valueless in some areas and there is an increase in bureaucratic control. There are attitudes of irresponsibility and no initiative. These are the order of the day to a great extent in our country, but even more in other countries. There is little concern for the needs and

lives of others, and there is a spirit of
lawlessness abroad.

Liberty and responsibility are opposite
sides of the same coin. At a very early
age children begin wanting liberty and
privileges—which the parent also wants
for the child—but training and discipline
are necessary first. Weaning is an early
training principle. The child learns how
to hold things, to walk, to ride a tricycle,
and the parents begin to define training
objectives which will permit the child to
take on responsibility. He is fed, and
then he wants to hold the spoon for himself. The little fellow wants to do everything for himself. You see, he wants liberties and privileges. However, wise parents do not give him those liberties until
he can also accept the responsibilities that
go with the privileges and liberties.

The training doesn't stop. Upon reaching adulthood, a person receives all the
privileges and liberties of adulthood, but
at the same time must assume the responsibilities that accompany them. Usually
there is some period of adjustment, and
occasionally it is critical. Those who do
not assume those responsibilities live un-
successful lives, and some even wind up in
prison.

Training is the method of external controls employed in maturing the child. His
increasing ability to assume an increasing

load of responsibility is a direct result of training and maturing. Initially, controls are from *without*. The purpose of training is to condition the child to *control himself,* and the result of proper training is a greater willingness to accept responsibility.

There is one element which develops during training and maturity, and it determines the level of independence—or conversely, of dependence. It determines the growth as well as the final responsibility and liberty that the individual will achieve. That element is character. Character development precedes responsibility, because it takes character to assume a degree of responsibility. The very essence of character, therefore, is the willingness to assume responsibility. Character precedes responsibility, which in turn precedes liberty. In other words, liberty is the assumption of responsibilities, which in turn is dependent upon character.

Character building, therefore, is training that is essential to equip the individual to assume responsibility, and this results in earned liberties. Character is a quality of life. As such, it affects personalities, attitudes, and life standards. Character is reflected in the individual's quality of conviction, objectives, and lifestyles. It is basic to his philosophy of life. This is one of the most profound truths you will ever consider. It is a significant principle of the American citizen. It is a fundamental truth and thesis of life.

That which is true abut character, responsibility, and liberty in regard to an *individual,* applies equally to a *nation.* For every liberty possessed by an individual or nation, there is a corresponding level of responsibility. The level of liberty which a nation achieves is equal to and cannot exceed the nation's level of character. The more responsible a nation becomes, the more liberty it assumes. The less responsible the people are, the less liberties they can retain. The more character a nation possesses, the more strength it has. The less a nation obeys God and the less character it possesses, the weaker it is. Thus, the weaker a nation becomes, the more it flees from its enemies. In every respect the loss of character results in the loss of liberty, whether individually or on a national level.

This means, then, that people *get* as much government as they *need*—as much government as the strength or weakness of their character requires. However, historically whenever character declines, that decline is paralleled by a growth in government. As the people decline in moral strength, the government takes over some freedom to replace that strength, so it can provide for the needs of the people. People accept and like the fact of security and dependence offered by the government. Thus, one result of the character decline is the growth of a totalitarian government. As character levels decline, the administration (bureaucracy) within the government begins to grow, and various agencies begin to appear, to perform more and more services for the people.

A nation void of character can neither demand nor expect good government. Good government is not a byproduct of a weak nation. The destiny of a nation is determined by the lives of its people. The quality of government declines with the quality of the people. The character of the government merely reflects the character of the people. Only if those people are virtuous can they demand their government be the same. If the people are righteous, they will elect righteous leadership. They will demand quality and high standards in their government. As the character of the people declines, the result is that less and less pressure is extended on their government to uphold basic principles. Therefore, there is a corresponding decline in the quality of government. And the monster we have created becomes "power mad" to the degree that it no longer feels it is necessary to respond to the requests (or demands) of those who placed it in control.

It has been truly said that the people get the kind of government they deserve. It is also true that the government will conform to the kind of character that is within the people. Government policies will simply reflect that character. You have heard of the "vicious cycle," well, consider *this* one: A nation that has less character than its people will ultimately lead them down into being a less disciplined people, and in turn, because they then require more control, will lead to a government that will *impose* more discipline upon them (for their own good, of course). The government begins to supply more security for the less responsible people (at the expense of those who are still trying to be responsible, needless to say), and security is an appeal to the lower nature of man, not an appeal to his character. Security, in fact, leads to a limitation of personal liberty. (Did you know that average employed Americans work until May or June of each year for the government before they begin to work for their own needs—if all our taxes were to be worked off first.)

When government provides increased security for the people and assumes more responsibility for the people, it will also begin to exert greater controls over the people. The government slowly assumes power over the people and eventually gains absolute power, becoming a totalitarian administration.

Without character, liberty cannot and will not survive. It is simply impossible. The people and the nation must have character.

There is an ultimate source of character, and that source is God's Word, the Holy Bible.

Today there is a great movement in America which *is* teaching character. One popular news magazine described it as the fastest growing movement in America today. That movement is the Christian school movement—education at the hands of the churches of America, education where the Bible is the premise and moral rules are practiced.

This movement of education today—the Christian school movement—has been challenged in this country, as have many of the other character traits of our land. Today many states have passed laws which attempt to limit private Christian institutions. Some states have passed laws stating that the State is the custodian of the child, and not the parent.

The Bible does not teach this. The *Bible* rather teaches that the center and focal point of all society is the home, and that the parents are the custodians of the child, charged by God with the duty of teaching them and developing their character.

Contrary to the Bible, *Socialism* teaches that the State is the custodian of the child.

The founding fathers of our nation, those who established this great Republic, believed that we should maintain freedom whereby the family unit, under God, was central. They believed in government of the people, by the people, and FOR the people, and our first school system taught this. However, across the land today laws are being passed in total accord with the socialistic trends of our day. Then the State, rather than the parent, teaches the child how to think and how to vote.

Only a short time ago I was talking to a Christian school leader and he said to me, "Reverend Lindsey, things are bad— our only hope is for a generation of young people trained in our schools."

I said, "Sir, God ordained the *church* as the basis for the best society."

This man replied to me, "Yes, Reverend Lindsey, but God never intended for the State to educate our children, but rather for our children to be educated in a Godly atmosphere. Unless the churches of America will take over the training of our children as part of their church Christian education, we will never see this nation great again."

I said, "Sir, will you please clarify that."

He said, "If God will give us one generation of graduates from our Christian
institutions, they can infiltrate State and
Federal government.   They could return
morals and principles to government, and
then we could spare this nation from its
final fall."   He went on to say that the
only fear he had was that God would not
give us one generation of graduates from
our Christian institutions.

I gave much thought and serious con-
sideration to that statement, and I must
concur with what he said. Morals, character, responsibility—they all go together.
You cannot have a *responsible* voting
populace without having a *charactered*
voting populace.   Therefore, the only way
for our nation to survive is to have responsible government, and honest government officials who will tell the American
people the truth.

Someone asked me in a meeting awhile
back, "Brother Lindsey, do you think
that the government officials in Washington actually know the truth about the
energy crisis?"

I answered, "Sir, we are faced with only two options: Either we have the most
disillusioned, misinformed government that
has ever existed, or our government officials are the biggest liars that ever walked
the face of the earth—it must be one of
the two."

I am persuaded after what I have seen with my own eyes and heard with my own ears, after two and a half years on the North Slope of Alaska, that there is no energy crisis. There has never been a true energy crisis, and there will never be a genuine energy crisis. There is a crisis, but it is an artificially induced crisis for the purpose of controlling the American people, and perhaps the rest of the world, as well. After all, if they control our energy and our communications systems— what else is left . . . *they control us!*

God grant that our nation will wake up and that some character will again fill the halls of our great institutions and government offices. God grant that once again private enterprise will be allowed to function and incentive will be a worthy motivation, so that we will once more produce as we have in the past, leading the world with our ingenuity, industry, intelligence, independence, and integrity. God grant that once again character will flow through our nation, and we will witness a great revival of truth and honesty sweeping across this country.

The only hope for America is for a rebuilding of character, which in turn will bring on the acceptance of responsibility. God grant this to come to pass, in individual homes, in positions in high places, and in government across this land.

As to what you can do, *be a diligent voting populace.* Make yourself aware of the issues and investigate the facts behind them. Don't be gullible or naive, swallowing all the propaganda your are spoon fed (by the media and other sources). Advise—in writing—your elected officials of the fact that your vote helped put them there, your taxes pay their salaries while there, and you will be carefully watching their performance while there. Tell them how you expect them to represent you on matters—especially those concerning individual liberties and free enterprise. (You can't expect them to read your mind.) Tell them how you feel about the so-called energy crisis and the self-appointed overprotective protectors of our environment who have tried to hinder the development of present energy sources and alternate sources for future energy needs. Stand up and be counted! . . . and if they ignore your input . . . *vote them out at the next election!* Just because they are the incumbent doesn't mean they are better qualified to do the job . . . in fact, if the incumbent wasn't representing you in accordance with your wishes anyway, how could you possibly be any worse off with someone new?

Let *your* character grow, and be diligent in passing on—by word and example—to your young people the character and instruction required to keep our nation great and free.

Then once more, *America will be great,* not only in international prestige, but more importantly, in moral and spiritual leadership.